TECHNOLOGY, INNOVATION and POLICY 8

Series of the Fraunhofer Institute
for Systems and Innovation Research (ISI)

W0107565

Gunter Lay · Philip Shapira
Jürgen Wengel (Eds.)

Innovation in Production

The Adoption and Impacts of New Manufacturing Concepts in German Industry

With 61 Figures
and 6 Tables

Physica-Verlag

A Springer-Verlag Company

Dr. Gunter Lay
Diplom-Sozialwirt Jürgen Wengel

Fraunhofer Institute for
Systems and Innovation Research (ISI)
Breslauer Str. 48
D-76139 Karlsruhe, Germany

Professor Dr. Philip Shapira
School of Public Policy
Georgia Institute of Technology
Atlanta, USA
and
Fraunhofer Institute for
Systems and Innovation Research (ISI)
Breslauer Str. 48
D-76139 Karlsruhe, Germany

ISBN-13: 978-3-7908-1140-7   e-ISBN-13: 978-3-642-99801-0
DOI: 10.1007/978-3-642-99801-0

Die Deutsche Bibliothek – CIP-Einheitsaufnahme
Innovation in production : the adoption and impacts of the new
manufacturing concepts in German industry; with 6 tables / Gunter
Lay ... (ed.). – Heidelberg; New York : Physica-Verl., 1999
(Technology, innovation, and policy; Vol. 8)
ISBN 3-7908-1140-8

Cover design: Erich Kirchner, Heidelberg
SPIN 10697689   88/2202-5 4 3 2 1 0-Printed on acid-free paper

Contents

Figures and Tables

LIST OF FIGURES

XI

XIII

LIST OF TABLES

Abbreviations

BMBF	German Ministry for Education, Science, Research and Technology
CAD	Computer-aided design
CAM	Computer-aided manufacturing
CIM	Computer-integrated manufacturing
CIP	Continuous improvement processes
CNC	Computer numerical control
DM	Deutschmark
EDP	Electronic data processing
EMAS	Environmental management audit system
EU	European Union
FZK	Forschungszentrum (Research Center) Karlsruhe
ISI	Fraunhofer Institute for Systems and Innovation Research
ISO 9000	International Organization for Standardization – procedures for verifying quality
ISO 14000	International Organization for Standardization – procedures for ensuring environmental compliance
IWH	Institut für Wirtschaftsforschung (Institute for Economic Research) Halle
JIT	Just-in-time (system of inventory control and industrial production management)
MRP	Material requirements planning
NC	Numerical control
PFT	Projektträger Fertigungstechnik (Program Administration Agency for Manufacturing Technology) at the Forschungszentrum (Research Center) Karlsruhe
PPC	Production planning and control
R&D	Research and development
SME	Small and medium-sized enterprise
TQM	Total quality management
TT	Technology transfer

Data Source

Unless otherwise indicated, the data cited in the text and in the tables and figures of this book is drawn from the Innovation in Manufacturing Survey of the German investment goods sector, conducted in 1995 by the Fraunhofer Institute for Systems and Innovation Research. The number of companies responding to this survey was 1,305. Further details about the survey methodology are contained in the appendix.

1 Introduction: Perspectives on German Industry and Its Competitiveness

Gunter Lay and Philip Shapira

1.1 German Industrial Competitiveness in Context

Many Germans look back to the 1970s as a highpoint in their nation's industrial development. The hallmark "Made in Germany" then signaled not only well-made products, but also a highly efficient industrial system. German industry was recognized internationally as a model of high achievement; the country's engineering, automotive and chemical industries were at the forefront of technology; and German products had an international reputation for quality and performance. At the time, customer demand for German products was so high that many manufacturers faced labor shortages as they tried to obtain workers to meet their order books.

However, by the 1980s, the situation began to change. Step by step, German industry seemed to lose competitiveness. This was symbolized in the American automotive market when, after vastly superceding Germany's previously dominant position in imported small cars, Japan's automotive producers then went on in the 1990s to match and, in the view of many consumers, outperform Germany in the prestige luxury vehicle segment. In other industries too, German firms found that rivals elsewhere could increasingly match their quality and design and better their cost and customer responsiveness. Commentators both inside and outside Germany began to suggest that German industry had lost competitiveness (see, for example, Dornbusch 1993, American Chamber of Commerce in Germany 1996, or Meil 1998).

Although there are diverse interpretations of Germany's industrial malaise, most discussions emphasize several common features. First of all, production in Germany is said to be too expensive. In particular, Germany has high industrial labor costs relative to its principal industrial competitors. In U.S. dollars at current exchange rates, total hourly compensation for German production workers averaged $31.87 in 1996, compared with $20.84 in Japan, $17.70 in the United States, and

$5.82 in Taiwan (U.S. Department of Labor 1998). Not surprisingly, it is often suggested that the hourly rates paid to German workers are much too high. Some businesses argue that competitive production is difficult compared not only with the newly industrialized countries but also against developed industrial economies. Additionally, working times in Germany have been reduced to among the lowest levels in the industrial world. German businesses complain that short weekly working times combined with long holiday periods make it even harder to maintain profitable production within the country.

Second, German enterprises are said to be too slow. For example, "time to market" for a product innovation is believed to be longer in Germany compared with Japan (see Womack, Jones and Roos 1990). The reason for this competitive disadvantage is usually said to be the inflexibility of firm structures in Germany, which are believed to be less capable to coping with today's shorter product life cycles. Clearly, if German enterprises develop their new products in the conventional way – where development tasks are done sequentially and thus slowly – they will enter the market behind competitors who adopt speedier, concurrent approaches to product development.

A third argument focuses on the huge number of laws and regulations that affect German businesses. This "over-regulation" is said to suppress innovation and entrepreneurship. While businesses and individuals are felt to be less restricted abroad, in Germany creativity is trapped within a tight net of rules. It is argued that a morass of barriers frustrates entrepreneurs with good ideas. Innovative concepts are given up or transferred to foreign countries where the business climate is more favorable. Promising innovations developed by German research institutes have only limited chances to be commercialized by German firms. This lack of innovative products is an important reason for falling behind.

A fourth argument states that German enterprises are too technology focused, leading them to overlook customer preferences. By training, culture, and tradition, German managers have a strong engineering orientation. In problem solving and in design, German engineers tend to pursue the most sophisticated and advanced concepts. In the development of products, it is suggested that these characteristics promote "techno-centric" solutions instead of the pragmatic, easy-to-use, and less costly products preferred by the mass of customers. In contrast, it is felt that foreign competitors apply technology as a means rather than an end, thus giving customers what they want.

A fifth factor commonly mentioned as contributing to declining German industrial competitiveness is the power of German trade unions and the influence they have over enterprises through legal rights to codetermination. It is said that codetermination makes it difficult German enterprises to react speedily and flexibly to changing markets due to the time required to reach agreements between management and labor representatives. Moreover, the compromises so reached often fall short of what the market requires.

Last but not least, the extensive promotion of environmental protection in Germany is said to be counterproductive to economic competitiveness. German public opinion is greatly concerned about environmental degradation, recycling and the conservation of nature, and businesses have to respect these attitudes and the environmental regulations and requirements that result. But, says industry, these pressures to address environmental concerns add further to the expense and delays associated with production in Germany.

The story of declining German competitiveness is bolstered by Germany's high unemployment. By the late 1990s, joblessness in Germany had exceeded 4 million, with an unemployment rate of more than 11 percent. Although German unemployment levels are comparable to those of France and Italy, the German unemployment rate is considerably higher than in the United States, Japan (despite its long-running recession), and other European economies such as Britain and the Netherlands. Moreover, the overall growth rate of the German economy has been weak for most of the 1990s, with lagging domestic demand for goods. Despite low interest rates, German firms have been reluctant to invest heavily at home. Much of East Germany – or what was the former German Democratic Republic – remains particularly economically depressed, giving lie to the hope that unification would quickly result in expanded markets and widespread prosperity. Such problems are symbolized in the falling value of the German Deutschmark (DM), which has lost significant value over the last decade compared to other currencies. A few years ago, the exchange rate was 1.50 DM to the US dollar. Nowadays, Germans have to pay 1.80 DM. The British pound and the Italian lira have also gained a stronger position compared to the DM. The weakness of the Deutschmark has fuelled German fears about entering the European monetary union.

Yet, despite the gloomy perspective of many current interpretations of what is happening to German competitiveness, other facts and trends present a somewhat different picture. For example, German productivity continues to be very high, so that even after high hourly wages, final wage costs *per unit* of output remain internationally competitive (aided, in part, by the lowered international value of the Deutschmark). Similarly, exports of German industrial goods have risen to new records. Exports per capita in Germany are more than twice as high as for the United States. Compared to the Japan, Germany goods exports are more than one third higher per capita. Undeniably, German goods still can compete on the world market.

It is also evident that major German companies are restructuring and restoring profitability. In recent years, strong profits have been registered at leading German firms, such as Bayer, Hoechst, BASF, Daimler-Benz and Volkswagen (see, for example, Der Spiegel, 1998, pp. 99). Moreover, a new profit-oriented management attitude is emerging in major German enterprises that focuses much more on shareholder value – an Anglo-American concept once thought to be foreign in German, but which is now increasingly accepted. Stock markets have reacted positively to this new attitude and to strong enterprise profitability, driving German stock market indices to new heights.

Finally, while unemployment is high in Germany and there have been jarring outbreaks of violence (particularly against immigrant workers) by disenchanted youth, the economic and social system continues to function at a high level. Although equity has yet to be achieved between the eastern and western parts of reunified Germany, prosperity is still evident across much of the country. Most German households have incomes and consumer goods to sustain comfortable living standards, while educational and health systems are maintained. The social security system provides relatively good benefits for the most severely affected people.

The point about these conflicting accounts is that it may be misleading to rely too much on a few examples, stories, or anecdotes about German industrial competitiveness. Nonetheless, it is apparent that the challenges of globalization to Germany's core manufacturing competencies have been felt forcefully by German industrialists. Industrial restructuring has occurred, sometimes on a massive scale – and at times with unfortunate consequences for German workers who have lost their jobs. Many German firms have expanded production overseas, in important markets such as the United States, as well as in lower-wage countries. New international strategic alliances have been formed. At the same time, existing manufacturing operations in Germany have been reorganized using new technologies and production concepts to improve productivity, quality, and flexibility – although, as contributors to this book will discuss, not all German firms have fully adopted the possible range of new production concepts.

An underlying rationale of this book is that the pace and direction of production change within German industry in response to globalization has not, to date, been widely recognized, let alone well documented. Companies have employed a variety of business and production strategies to address the kinds of concerns about German competitiveness highlighted earlier. Throughout industry, major efforts have been pursued to maintain unit wage costs at competitive levels, speed up development cycles, strengthen innovation, and improve customer relationships. These efforts have proceeded in ways that have mostly respected, rather than run roughshod, over existing structures of employee involvement and environmental protection. Of course, the pace of change is not uniform: there are clearly leaders and laggards in the re-orientation of German industry to new competitive conditions. Moreover, these efforts have not necessarily resulted in all factories and jobs being retained. But – as we will see in later chapters of this book – there is evidence that those firms that have tried to sustain competitiveness through improving the performance and innovativeness of their German-based operations have generally fared better than firms whose principal strategies have mainly involved cutting costs or relocating production to other countries.

This book is an examination of the take-up and implementation of innovation within production systems by German manufacturers in recent years in response to growing competitive pressures. The book focuses on Germany's investment goods sector – which includes the engineering, machinery, electrical, automotive, and electronic firms at the core of the country's manufacturing base. The contributing authors to the book analyze the specific (and diverse) ways through which firms in this sector are modifying, if not restructuring, the ways in which they manufacture, through interrelated technological, organizational, workforce, and, in some cases, locational changes. The effects on such factors as productivity, innovation, and employment are examined too, along with implications for business planning and policy. The analysis draws on data reported by the more than 1,300 manufacturing establishments in the investment goods sector who participated in a wide-ranging 1995 survey of innovation in manufacturing and production undertaken by the Fraunhofer Institute for Innovation and Systems Research (ISI).

The ISI survey provides a unique base of information about the diffusion of new manufacturing technologies and new organizational concepts in the German investment goods sector (see following section for a discussion of the composition of this sector). Also examined are the relationships between technological and organizational innovation, corporate production strategy, and business and employment outcomes. (Methodological details about the survey are contained in the appendix to this book.)

The special survey approach used in this book is critical in grasping the complexity, variety, and differential outcomes of change within German industry. The macroeconomic data published by national statistics offices tend to report averages, meaning that underlying variations are often masked. Moreover, if only macro economic data is used, it is impossible to detect causalities between structures and effects. Similarly, individual case studies of corporate response are helpful in providing vivid insights into particular situations, but it is too much to expect a valid analysis of a whole sector through cases alone. However, supported by these other methods, ISI's sector survey allows the possibility to link structural changes within manufacturers to their economic impacts on firms and workers. This allows the authors to examine how German industry is responding to global competition using new production approaches and under what conditions those strategies are likely to be more or less successful. This is not an entirely new question – see for example the studies of U.S. competitiveness by Dertouzos, Lester, and Solow (1989) or of Hong Kong by Berger and Lester (1997). However, the ISI study does represent one of the first opportunities to ask such questions in the German context.

The next section discusses the German investment goods sector and its key characteristics. This is followed by an overview of the book.

1.2 The German Investment Goods Sector

"Investment goods" are typically products that are durable, complex, and which, in their final form, often are significant purchases that are then employed by end users to further their economic aims. Investment goods include machine tools, engineering and electrical products, automobiles, aircraft, and computers (see the appendix for a detailed definition of the sector). A common feature of the sector is that metalworking is often necessary. High levels of design and the production and integration of complex parts are also frequent. Thus, machinery, work structures, and employee qualifications are similar across the sector, especially when compared with other industries such as chemicals, basic steel production, or food processing. The similarities found within the investment goods sector make it feasible to study through common survey procedures.

The investment goods sector is the foundation of German manufacturing industry. Investment goods producing firms employ more than one half of the people working in German manufacturing and about 15 percent of all workers across all sectors of the German economy (figure 1-1).

The significance of the investment goods sector is illustrated by its large role in German exports. In 1996, 56.8 percent of German exports (by volume) consisted of investment goods. This percentage far exceeded the share of the other sectors. The share of German exports contributed by other sectors was 13.1 percent for the chemical industry, 4.3 percent for food producers, and 2.4 percent for the plastics sector.

During the 1980s, the German investment goods sector performed strongly, reaching a peak in 1991. Between 1980 and 1991, the German investment goods sector increased its gross value added (at constant prices) continuously from DM 250 billion to DM 350 billion (figure 1-2). Over the same period, the number of employees grew from 3.5 million to 3.9 million (figure 1-3). Value added rose about 3.5 times faster than employment, indicating substantial improvements in productivity.

Figure 1-1   Employment in Germany, by economic sectors, 1996

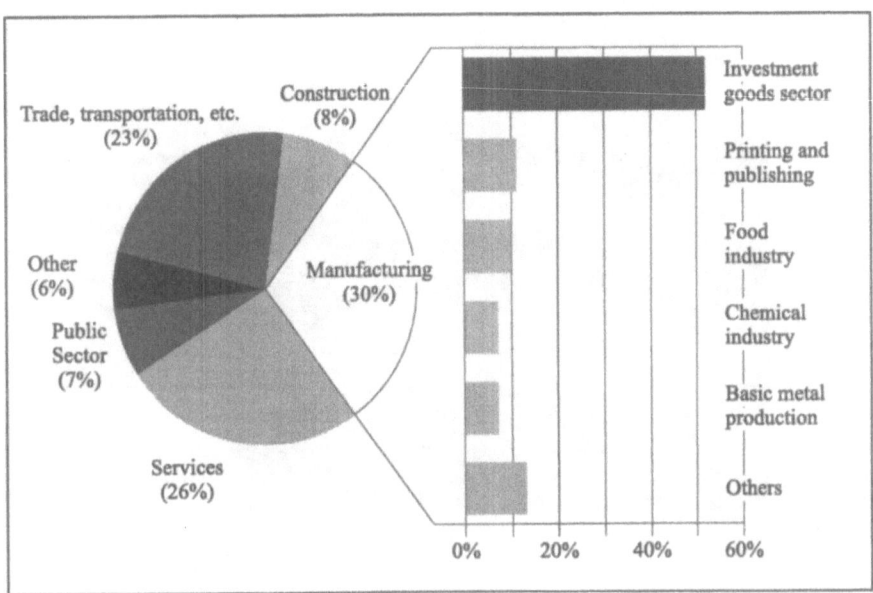

Source: Federal Statistical Office 1997

Figure 1-2   Gross value added in the German investment goods sector,
                  1980-1996

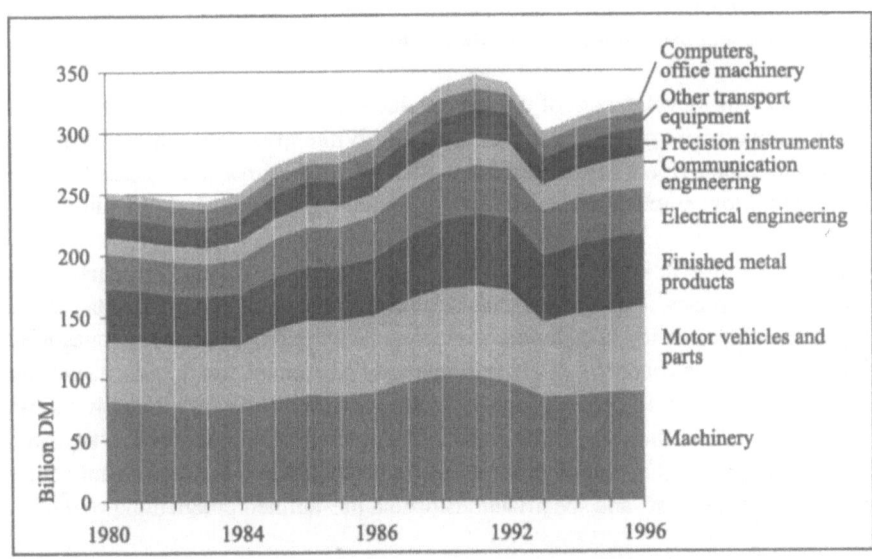

Data 1991 prices, for old Länder of West Germany. Source: DIW 1997

Figure 1-3   Employment change in the German investment goods sector,
               1980-1996

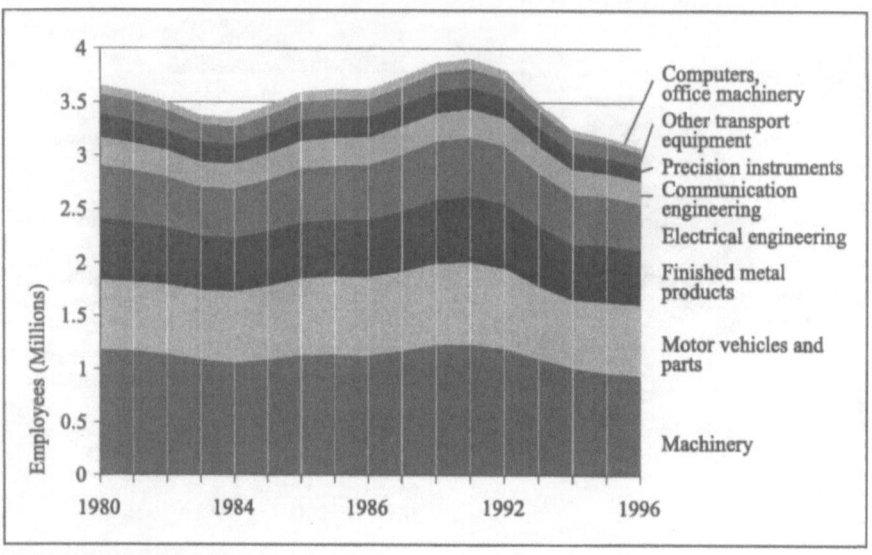

Data for old Länder of West Germany. Source: DIW 1997.

However, in the early nineties there was a sharp decrease in the sector's gross value added. Between 1991 and 1993, German investment goods companies had to reduce their production by nearly 15 percent. Due to this reduction, they discharged more than 10 percent of their employees. The employment in this sector of German industry was cut back to the level of the beginning eighties.

Since 1993, the development of the gross value added and of the employment in the German investment goods sector has diverged. While gross value added is increasing again, employment has continued to fall. In 1996, gross value added reached DM 325 billion; employment, however, has fallen beyond a level of 3.1 million.

This jobless growth could be the result of a new kind of production paradigm in German enterprises. Thus the ISI manufacturing innovation survey had its focus in asking for the diffusion of new manufacturing technologies and new organizational concepts, for qualifications and deployment of personnel, for trends in working time and remuneration concepts and for economic indicators. Additionally the survey sought information about the establishment (number of employees, turnover), its industry group, the kind of products and whether the products are manufactured to customers' requests and specifications or to a pre-defined production program.

1.3   Overview of *Innovation in Production*

As the previous section showed, the investment goods sector is critical to the German economy. But how have companies in this sector adjusted their production strategies in response to new competitive pressures? This is a core theme of this book, as it considers how companies are changing their manufacturing processes through the adoption of new production concepts. These concepts include the introduction of new organization principles, the development of quality management and redesign of supply chains and customer links. Although new "hard" technologies, such as computer-aided machines or computer networks, are often involved, a characteristic of the current generation of new production concepts is their organizational emphasis. The chapters in the book define more specifically the constituent elements of new production concepts and examine their diffusion. Particular chapters also consider the impacts of new production concepts on productivity, innovation, and employment, and analyze implications for business strategy and industrial policy. All chapters draw on the empirical base provided by ISI's survey of the German investment goods sector.

The tension between the increasing turbulence of market and production conditions and traditional business structures provides the underlying theme of the chapter on the diffusion of new production concepts in Germany by Kinkel and Wengel. The authors argue that there is now a consensus on the principles that should guide firms as they adapt their structure and organization to match a fast changing business environmental. New production concepts are the means through which these principles are put into action. Kinkel and Wengel then go on to discuss the diffusion of a series of specific new production concepts in the German investment goods industry. They focus on the take-up of new methods of work organization and management, the improvement of quality, and the reshaping of supplier and customer relationships. Kinkel and Wengel observe that large companies are quicker at introducing new production concepts than smaller ones, but also note differences in adoption by industrial branch and type of production. The authors consider the adoption of new production concepts in terms of their appropriateness – an important qualification since not all methods are appropriate to all companies in the sector. Nonetheless, the conclusion is that relatively few companies have comprehensively adopted new production methods. Kinkel and Wengel argue that concern about cost-reduction in Germany has overshadowed the potential role that performance-enhancing modernization and improvement strategies.

The preoccupation with cost-reduction in German industry is not surprising given Germany's high labor costs. Social expenses charged to employers and employees represent a significant part of hourly compensation costs in Germany. There is ongoing debate in Germany about curtailing these social charges, reducing the benefits they provide, and improving the efficiency of the public services they support.

Meanwhile, in an effort to cut costs, many German firms have reduced employment at home and shifted elements of production to less costly locations. Yet, from a German perspective, such strategies have limitations. Efforts to change the structure of labor costs and the provision of social services face many political obstacles and progress is slow, while the loss of jobs within Germany is clearly an unpalatable solution.

Thus, the essential challenge facing Germany is one of maintaining international market competitiveness yet at the same time securing (if not creating) domestic employment while sustaining (rather than reducing) the high economic and social living standards that the population desires. From an industrial perspective, three interrelated production approaches are fundamental to meeting this challenge. First, to offset high production costs at home, German firms need to increase productivity. Second, German firms must also upgrade the value they deliver to customers by improving quality and timeliness. Third, German firms have to introduce innovative products that can expand their market position. A core argument in the opening chapters of this book is that all three approaches are aided by the use of new production concepts. In chapter 3, Lay, Kinkel, and Dreher focus on the first two approaches. They find that firms in the German investment goods sector that have adopted new production concepts such as teamwork, decentralized decision-making, non-buffered materials processing, or just-in-time delivery report higher productivity (measured by value-added per employee) than comparable non-adopters. Use of new production concepts is also associated with higher quality and reduced materials stocks. However, different types of firms vary in the results obtained from using new production concepts. For example, in the mechanical engineering sector, the benefits associated with the adoption of teamwork are greater in smaller firms producing more routine items than in medium-sized companies using specialized expertise to make customized products. The authors conclude that firms need to carefully select which particular new production concepts are most appropriate to their specific business situation and competitive strategy.

In the following chapter, Lay places emphasis on the third strategy – the introduction of innovative products that can expand market position. Lay finds that in German investment goods firms where innovative products comprise a high proportion of sales, there is also a faster sales growth rate. He then goes on to examine the connections between product innovation and the adoption of four new production concepts in the German investment goods industry: simultaneous engineering, interdepartmental development teams, research and development collaboration with suppliers and customers, and continuous improvement. Although there are important variations by different types of firms and product sub-sectors, Lay identifies general relationships between the use of these new production concepts and corporate performance in product development and innovation. Interestingly, firms that were initially furthest behind in innovation performance demonstrated the greatest gains from adopting new innovation-facilitating production concepts.

With joblessness in Germany currently running at an extremely high rate, the impact of new production concepts on employment is a thematic area of great interest and concern. The next three chapters consider particular issues within the employment theme. A foremost issue is how the spread of new production concepts affects employment levels in industry. When companies change their production strategies and introduce new technologies and ways of working, there is frequently a fear that reduced employment requirements may result. On the other hand, it is often suggested that the use of new production concepts in industry can promote job growth. In chapter 5, Lay argues that the critical factor in determining the direction of associated employment effects is not the adoption of new production concepts *per se*, but the strategic context in which those concepts are deployed. Generally, there is not a significant relationship between the adoption of new production concepts and employment change, based on evidence from Germany's investment goods sector. However, the adoption of new production concepts is associated with positive employment growth in firms that pursue a business orientation emphasizing improved performance and quality as opposed to straightforward cost-reduction. In a link with the findings on innovation and market expansion noted in chapter 4, Lay explains that a performance and quality orientation encourages new production concepts to be deployed in ways that further bolster market success – and the related growth in output leads to greater employment requirements. Conversely, firms that emphasize cost-reduction are more likely to use new production concepts to eliminate the expense of labor. Lay recommends that more attention be paid to encouraging firms to use new production concepts in support of performance-focused business strategies.

New production concepts can also be expected to affect the organization and structure of employment in industry. Again, context and strategy is critical. In chapter 6, Wengel and Wallmeier review the systems for employee co-determination that have been established in modern German industry. New production concepts, particularly those that promote greater employee participation, teamwork, or decentralized decision-making, are implemented in the context of the existing systems of employee co-determination. There is some apprehension in Germany, particularly among managers, that further increases in employee participation will negatively affect business competitiveness, responsiveness, and flexibility. Wengel and Wallmeier find to the contrary – that greater use of participatory production approaches in the investment goods sector is associated with improved business performance. In the subsequent chapter, Lay and Mies report that, where implemented, the introduction of flexible working hours can also lead to business benefits. However, the authors of both chapters observe that the diffusion of participatory production methods and flexible working hours in the investment goods sector is still patchy and incomplete, suggesting that there are opportunities for greater progress on these fronts.

The expansion of manufacturing activities by German companies in locations outside of the country must be understood in terms of the broader globalization of production. Lower wage and regulatory costs in countries outside Germany have an influence on location decisions. In addition, there is a need to serve customers outside Germany and to tap new markets, capabilities, and opportunities in other countries and regional trading blocs. The growth of communications and transportation technologies has facilitated the location of production abroad. Sub-sectors within the German investment goods industry, particularly those that focus on high-volume and mass-production such as automobile assembly, have long been extensively engaged in international production. In other parts of the industry, for example in machine building, the more customized nature of production, requirements for investment in plant and equipment, and needs for skilled labor have tended to make firms arguably less footloose in the search for cheaper labor. However, demands on all sections of the German investment goods industry to internationalize production have increased as manufacturing costs in Germany has grown, markets in other parts of the world have expanded, and there has been a greater emphasis on management methods that encourage close relationships between investment goods suppliers and customers.

In chapter 8, Kinkel examines the transfer of production to locations outside Germany and explores the relationships between the relocation of production and the use of new production concepts. He finds that while large firms have led in the transfer of production to date, in coming years many more small and medium-sized German firms plan to relocate production abroad. Similarly, the planned rate of production transfers in engineering and in complex products is expected to match that found in the automobile industry and other mass-produced items. As might be expected, firms following a cost-oriented business strategy are rather more likely to have transferred production outside of Germany than firms that are pursuing a performance-oriented path. Kinkel relates the locational profile of firms to their production proficiencies. For high-volume simple items, lead times and development cycles are not much different when firms that have relocated production are compared with those that have not. However, for complex batch products, firms that have relocated production have a poorer performance on these parameters than firms with no relocation. Kinkel then goes on to define four categories of firms within the investment goods sector, based on their use of new production concepts and whether they have relocated production outside of Germany. He finds some evidence for the proposition that productivity, measured by value-added per employee, is highest among firms that have adopted new production concepts but have not relocated production abroad. Kinkel argues that this illustrates that Germany is still viable as a production location, as long as firms modernize their production through new technologies and organization methods.

The diffusion of specific production concepts is examined in the next two chapters of the book. One of the new technologies that German equipment suppliers are adopting in an effort to cut costs and improve value-added to customers is "tele-service". This allows vendors to remotely monitor the operation and performance of equipment installed in the factories of customers and to electronically diagnose problems and initiate maintenance. Dreher, Lay, and Michler examine the implementation of teleservice in chapter 9. About one quarter of German investment goods firms currently use teleservice, with the highest diffusion rates found among larger firms. The diffusion of this technology is expected to grow further in future years. The use of teleservice does not substitute for a firm's own maintenance staff, but it does offer users greater information and supplier support to complement internal capabilities.

Germany has the reputation of imposing a tough environmental protection regime on its manufacturing industries. Although industrialists raise concerns about the expense of compliance, environmentalists argue that environmental conservation is ultimately in the self-interest of industry, by reducing waste and opening new markets for environmental protection technologies and "greener" products. Chapter 10 considers the debate about the benefits and costs of environmental protection in the context of the discussion about new production concepts in industry. In this chapter, Fleig examines the adoption and impacts of two environmental production concepts in the German investment goods industry: dry processing (machining without lubricants) and environmental auditing. About one-fifth of German companies in the investment goods industry use dry processing technologies, but this seems to be a technology whose diffusion rate has peaked as few additional companies see the need or opportunity to adopt it in the future. For environmental audits, only about 6 percent of all investment goods companies are adopters, with the adoption rate higher for large firms and rather lower for smaller ones.

Fleig's underlying observation is that environmental investment in the investment goods industry is weak, and the take-up of measures is low. For dry processing, technical factors limit its wider applicability. But the diffusion rate for environmental auditing is surprisingly low, given Germany's vaunted environmental image. However, an additional 25 percent of firms plan to implement environmental audits in the future. Fleig finds little evidence that the adoption of either of the two environmental measures negatively affects profitability. Indeed, he hints that firms that lead in adopting environmental measures are more likely to be innovative and have a slightly better quality performance than non-adopters. Yet, these relationships are not very strong in the data reported by firms. While there are certainly measurement problems in identifying the full range of impacts from environmental measures, it is also plausible to suggest that most firms continue to view and implement new production concepts and environmental protection separately. Potential, mutually reinforcing, benefits are thus only weakly realized.

In the next two chapters, attention is centered on East Germany. The dramatic reunification of Germany in October 1990 abruptly brought together two economic and industrial systems that had developed, over the post-world war II decades, under radically different conditions. The state-sponsored industries of East Germany had to rapidly restructure to adapt to private ownership, market forces, the one-for-one conversion of the old East German currency to the Deutschmark, and the precipitous collapse of demand from Russia and other former Soviet-bloc economies. Significant parts of the old east German industrial base could not make this transition, leading to plant closures and traumatic job losses. At the same time, industries in West Germany were challenged to participate in the often costly reshaping of manufacturing units in the new German Länder (states), while simultaneously under great pressure to reengineer themselves to maintain competitiveness in global markets. It was visibly apparent that East German firms had to develop – or have developed for them – new business, management, and marketing strategies. In addition, manufacturing methods needed to be upgraded to improve industrial efficiency, since East German productivity was much poorer than found in the West. As Lay observes, in his analysis of the adoption of new production concepts in East Germany (chapter 11), East German investment goods manufacturers then lagged substantially behind their West German counterparts in using modern methods such as computer-aided design, computer numerical control, computer-integrated manufacturing, and product planning and control systems. It was thus hoped that investment in new advanced technologies, aided in part by federal subsidies, raise the productivity of firms in the new Länder to national and international levels.

By the mid-1990s, it was apparent that considerable technological investment had taken place in the surviving parts of the East German industrial base. Indeed, for three of the hallmark technologies examined by Lay, diffusion rates for the new and old Länder were comparable in 1995, while for the fourth, East German manufacturers had almost caught up. Yet, for investment goods manufacturers, East German productivity is still about over one-half of the level of West Germany. Lay probes why this is so. He examines several "measurable" factors that might account for East German poor performance. These include weak market demand and consequent over-capacity, high non-production overheads, lower density of technology use, weaker use of complementary organizational methods such as teamwork or just-in-time production, and differences in the complexity of production. With some exception for the last factor, little evidence if found for these possible explanations. Instead, Lay suggests that "intangible" elements are more significant. In particular, he theorizes that while investment in modern technology has accelerated in East Germany over the most recent decade, management, and workforce capability to take advantage of these technologies has not developed commensurately. Indeed, he proposes that a slower pace in adding new technological assets could result in faster productivity growth if complemented by the introduction of supporting organizational changes, skills upgrading, and management training.

Unfortunately, at least in the early 1990s, this was not a message that federal policymakers and the industrial constituencies of the new Länder wanted (or were able) to listen too. There was already growing evidence, from experience in West Germany and other industrialized economies, that advanced computerized manufacturing technologies were not so easy to use effectively and would not automatically result in significant productivity or quality improvements unless parallel improvements in training, organization, and management were made. Nonetheless, a program to promote computer-integrated manufacturing (CIM) was mounted in East Germany. Reinforced by the engineering orientation of most East German managers, there was a strong belief that closing the "technological gap" with the West would raise East German productivity and competitiveness. The CIM program for the new Länder was a manifestation of this belief, with some DM 100 million (about $56 million) in federal research ministry subsidies dispensed to more than 400 East German production equipment manufacturers in 1991 and 1992.

In chapter 12, Wengel examines the penetration, implementation, and impacts of the program to promote CIM in the new Länder. Wengel calculates that about one-third of eligible manufacturers were assisted, although he also finds a selection bias. Although many small and mid-sized firms were reached, larger and more technologically-advanced East German manufacturers were more likely to be attracted to the program and these participating units were more frequently owned by West German firms or by the Treuhand privatization agency. These firms used the federal subsidies as cost shares for new CIM hardware and software installations and for allowable training and other business management projects. Wengel observes that the CIM projects did influence management behavior and project decision-making. There was extensive use of outside consultants, and very few companies reported that they would have undertaken their projects in the same way without the program. For some CIM-related technologies, the adoption rate in the beginning of the program was lower for participating firms than for non-participants, although by the end of the program promoted firms had higher levels of CIM use than found in non-participants. Ironically, one of the "benefits" of the CIM program may have been to provide the time and cash for a more thoughtful strategy of introducing new technologies. However, by 1995, there was little difference between participating and non-participating firms in terms of economic performance (although CIM participants did have more optimistic views about the future). Certainly, the far-reaching economic, industrial and social changes underway in East Germany during the period affected firms across the board in ways that far outweighed the impacts of the CIM program. However, Wengel also concludes that, for most CIM projects, there was inadequate medium and long-term corporate planning to guide the use of new technologies and to ensure these investments were both necessary and fully effective.

The analyses of the East German experience are consistent with one of the major overall themes of the book: that upgrading manufacturing performance requires broader changes in management, training, and organization as well as new technology. In many instances, German firms, with their strong technical traditions, have had to rediscover by themselves the importance of "soft factors". The contributions to the book also moderate and qualify several commonly repeated views about the German industrial situation. Wage costs are high, but if coupled with comprehensive measures to upgrade production systems, increased productivity can be achieved and competitiveness maintained, even when manufacturing within Germany. Concerns about working hours, worker involvement in decision-making and environmental regulation are evident, but it turns out that the barriers to making change in production are not so great as popularly assumed. Finally, it seems clear that the biggest gains to adopting new technological and organizational innovations in production accrue to those firms that make these changes within a broader strategy of performance enhancement, as opposed to simple cost-cutting.

These are some of the principal conclusions of the book. The concluding chapter (chapter 13) reviews these conclusions more fully. The chapter considers the relationships between the use of new production concepts and business strategy – a critical issues since many German firms still emphasize cost-reduction as their principal approach to maintaining competitiveness. Insights from the various contributions to the book are also used to reinterpret some of the realities and myths of the current German industrial situation. This is followed by a discussion of implications for business management, policymaking, and future studies of technological and organizational innovation.

1.4 Bibliography

American Chamber of Commerce in Germany (1996), Studien zum Standort Deutschland, Frankfurt.

Berger, S., and Lester, R.K. (1997), *Made by Hong Kong*, Oxford University Press, Oxford and New York.

Der Spiegel (1998), Die kühnen Pläne des VW-Chefs Ferdinand Piech, 11, pp. 98-100.

Dornbusch, R. (1993), The End of the German Miracle, *Journal of Economic Literature*, 31, pp. 881-885.

Dertouzos, M.L., Lester, R.K., and Solow, R.M. (1989), *Made in America*, The MIT Press, Cambridge, Mass.

Meil, P. (1998), Der Blick von außen (View from outside). In: B. Lutz (Ed.), *Zukunftsperspektiven industrieller Produktion*, Campus Verlag, München, pp. 11-44.

Pond, E. (1998), Der zynische Blick auf Europa, *Die Zeit*, 10, pp. 3.

Womack, J.P., Jones, D.T., and Roos, D. (1990), *The Machine That Changed the World*, Rawson Associates, New York.

2 The Diffusion of New Production Concepts in Germany

Steffen Kinkel and Jürgen Wengel

2.1 Introduction

In the past, German industry was synonymous with high-performance products, high quality, success in international markets, and consistent growth. Today, the matter is more complex: while some German companies are very profitable and have improved their export market positions, other German firms have fared less well. The German media has reported many cases where industrial workers have been laid off, long-established companies have collapsed, or production has been transferred abroad. German enterprises can no longer count on the stable business conditions found in earlier periods. Instead, market and production conditions have today become more dynamic and increasingly unforeseeable. This "turbulent environment" is characterized as follows (Lay and Mies 1997):

- Market structures are changing. Tendencies towards saturation and shifts in demand structure are apparent. At the same time, there is a trend towards globalization of competition on all markets.

- Changes in customer demand are short term and unsteady. The number of customer-specific products and variants is continuously increasing.

- Customers demand high quality, reliability, and flexibility – and, at the same time, seek competitive prices.

- The rapid pace of technological changes, coupled with intense market competition, generate a need for high innovation rates and short product life cycles.

Germany's traditional structures of production and organization were not created for this turbulent business environment. On the contrary, the systems and management techniques traditionally employed by German companies are oriented towards functioning in a basically stable and forecastable environment. However, enterprises now have to adapt to the new business conditions – or else their survival is at risk. This has led to a great deal of interest in new production concepts that aim to enhance competitiveness through a fundamental reorganization of enterprise strategy, innovation, management, and organization.

Attention to new production concepts was first raised in Germany through the idea of "lean production". In the late 1980s, MIT's study of the international automobile industry highlighted weaknesses of the German system, particularly in comparison with that of Japan (Womack, Jones and Roos 1989). The message about the great importance of improving the organization of manufacturing enterprises was well received within German industry, but such ideas were sometimes misinterpreted as presenting simple remedies for deeper and more intractable problems. This was particularly noticeable for the stream of related concepts that were introduced in the 1990s to guide the reorganization of business, including "business reengineering" (Hammer and Champy 1993), the "agile enterprise" (Goldmann et al. 1995) and the "fractal factory" (Warnecke 1992). The perception of these new concepts as easily applied "solutions", together with the different messages each carried as to business strategy, led to confusion. The terms used to characterize new production concepts became fashion fads without inadequate content and with an ever-shorter shelf life.

Perhaps due at least in part to the misunderstanding generated by this array of competing ideas, discussion about new production concepts has been overshadowed by a purely cost-oriented debate about Germany as a location for industry. We suggest that an emphasis solely on cost may be a mistake. The prudent and company-specific implementation of the individual elements of new production concepts continues to offer significant opportunities to sustain industrial production and maintain jobs and wages within Germany. As the manufacturing innovation survey of the Fraunhofer Institute for Systems and Innovation Research (ISI) shows, the productivity of companies can be raised considerably by employing mutually consistent new production concepts (see Chapter 3).

In this chapter, the aim is to define and explain the elements of new production concepts and to consider the diffusion of these concepts within the German investment goods industry. There is an analysis of types of manufacturers by the extent to which they lead or lag in the use of new production concepts. This chapter thus provides a foundation for subsequent chapters to pursue more detailed analyses of the impacts associated with new production concepts. The chapter begins by outlining the key principles of new production concepts.

2.2 Principles of New Production Concepts

What is involved in the idea of "new production concepts?" The essential point is that the restoration and enhancement of industrial competitiveness in today's business environment calls for strategic, managerial, organizational and technical changes in the way manufacturing enterprises operate. Although analysts may differ on specific details, a basic consensus has emerged on the key principles that are involved (Dreher et al. 1995):

(1) Enterprises must meet the increasingly complex requirements of the market by simplifying their strategic and operational planning and management systems. This *principle of simplification* involves *reducing the complexity* of the product (by concentrating on its usefulness to the customer), of production (by concentrating on high performance process steps with a high value added) and of external interfaces (by reducing the numbers of customers and suppliers). When attempting to control their internal complexity, enterprises need to rethink their hierarchical structures and decentralize their decision-making processes. Achieving this often requires a shift of competencies, through autonomous responsibility and self-organization, to decentralized organizational units.

(2) *External and internal customer orientation* must be explicitly included in the strategy of the firm: close contact with external customers is regarded as the most important sensor for success in relevant markets. Moreover, within the enterprise, successive organizational units along the process chains should be regarded as internal customers. This requires the integration of plans as well as functions, thus enabling modifications in a firm's performance to be directly linked to external and internal market signals.

(3) The principle of *concentrating on value added* implies that inefficiencies should be avoided by confining the firm's activities to specific core activities. In order to do so, the scope of the enterprise's performance has to be optimized. Therefore, growing importance is given to the quality of the contacts with associated partners.

(4) In every part of the enterprise, consideration must be given to *communication and transparency* as a principle of openness in the flow and exchange of information. This includes intensive communication with customers in order to be able to identify their current requirements and also internal communication which aims at establishing short feedback and management loops within the decentralized units. In addition to openness about current actual performance, transparency about future business plans is very important in enabling decentralized management.

(5) The firm must support the ability, desire, and willingness of its personnel to work. Thus, *people as the main resource* of an enterprise is now a focal point, with employees regarded as primary contributors to improved performance rather than simply as a cost factor.

(6) The demand for greater flexibility and rapid customer response necessitates *an integrated view of the product and the production process*. In concrete terms, this implies an object-oriented formation of organizational units, instead of the functional orientation that has thus far been common. Planning and development

processes have to be shortened by introducing parallel steps so that faster and far-reaching innovations become possible. The social dimension involves bringing employees from various fields of work together in task-oriented project teams.

(7) Besides improvements through far-reaching innovations, improvement in small steps (*continuous improvement*) is a main principle of new production concepts. Thus, it is important to involve the skills and creativity of all employees on all levels. In this way, the enterprise can become a learning organization through constant feedback between suggested improvements and their effects on processes and procedures within the firm.

2.3   Implementation of New Production Concepts

The principles discussed above have generated a series of specific elements and measures that firms have implemented as they have adapted to competition in today's turbulent business environment. To what extent have manufacturers in Germany's investment goods sector actually implemented these elements?

Drawing on the 1995 Fraunhofer manufacturing innovation survey, we assessed progress along fourteen specific elements of new production concepts (see appendix for further information on the survey). These elements were organized into three major categories: new principles of work organization and personnel management, innovative quality management, and reshaping of the value added chain. For each element, we asked survey respondents a two-part question, first about whether they had *any experience* in using an element and, second, whether they had *comprehensively* implemented that element throughout their facilities (Table 2-1).

The results from the survey reveal a differentiated picture of the modernization of production structures in the investment goods industry. However, the diffusion of individual elements of the new production concepts is not as high as might be expected considering the intensive public debate on this topic over the last decade. This is particularly evident when the data on comprehensive use is considered, where only one element (ISO 9000) has diffused significantly – and then to only one third of establishments. Taken as a whole, Germany's investment goods firms still have a considerable way to go making necessary organizational changes to fully implement new production concepts.

Table 2-1   Diffusion of new production concepts in the German investment goods sector, 1995

New production concepts		Deployment level	
Major categories	Specific elements	Any experience Percent	Compre-hensive use Percent
(1) New organizational principles	Task integration	43	9
	Development teams	42	6
	Decentralization	24	8
	Team work	32	6
(2) Innovative quality management	CIP	44	18
	Quality circle	40	13
	Certificates (ISO 9000)	39	32
	No control of incoming goods	19	6
	Environmental audit	6	1
(3) Redesign of the value added chain	Manufacturing segmentation	40	11
	JIT-supply to customer	29	6
	JIT-supply	26	5
	Supplier concentration	26	4
	Kanban systems	19	5

In detail, the following picture emerges for the three categories of new production concepts examined in the survey:

(1)   New Principles of Work Organization and Personnel Management

- *Task integration* includes reintegrating and entrusting those activities to an employee that have become split up by an excessive division of work, for instance, the combination of production and testing tasks. This element of work organization is the most widely dispersed, applying to 43 percent of the manufacturers. However, the integration process has only been fully completed in one-fifth of these establishments, meaning that task integration has only been extensively introduced in less than 10 percent of establishments.

- The formation of temporary, interdepartmental *teams for product development within a limited time* is also widespread, applying to 42 percent of the manufacturers. A comprehensive, corporate implementation, however, is also comparatively rare, applying to 6 percent of all manufacturers.

- However, *decentralization* of decisions by transferring them to the level and site of action, with the aim to improve the basis for decision-making and to accelerate decision-making processes, has only been implemented by a little less than one fourth of the establishments. One third of these, i.e. 8 percent, state that they have adopted this element comprehensively.

- The realization of *teamwork* is still being discussed intensively. As a result, many manufacturers have yet to implement teams in their production and office units. According to the results of the survey, the solutions in about one third of the manufacturers correspond to a more discriminating definition of teamwork with revolving and integrated directing as well as quality control tasks. Pilot tests or applications in some sub-units are more frequent. Only about every twentieth manufacturer revealed a comprehensive use.

(2)   Innovative Quality Management

- The principle of *continuous improvement (CIP), quality circles,* and *certification in compliance with ISO 9000* as elements of an innovative quality management not only show the highest diffusion, but also have the highest diffusion within manufacturers. Measures for implementing *continuous improvement processes (CIP)* are the most widespread element, having been introduced by 44 percent of the establishments.

- By *certification in compliance with the ISO 9000* series of standards, an establishment can show that it has organizational structures that make it probable to achieve a certain standard of quality. Certification is quite widely dispersed, having been obtained by nearly 40 percent of establishments, and is usually implemented comprehensively.

- However, the ISO certificate as proof of innovative quality management methods should not be overestimated. The indicator *no control of incoming goods* indicates that up to now probably not more than one fifth of the capital goods manufacturers have really tackled very far-reaching integrated approaches to total quality management (TQM). And it is only applied comprehensively in 6 percent of the establishments.

- Thus far, 6 percent of the manufacturers have adopted an *environmental audit*, which, however, was usually limited to only a few areas of their activities. This is still a relatively new issue, but one which has been the topic of very intensive debate (see also chapter 10).

(3)   Redesign of the Value Added Chain

- An important measure in redesigning the value added chain is *manufacturing segmentation*. This involves the transformation from a volume production arrangement (for example, linear mass production) to an arrangement oriented on product lines (often implemented through reorganized production cells or work clusters). Manufacturing segmentation has a relatively high diffusion, amounting to 40 percent. But only about 11 percent of establishments have, as of yet, completed this process comprehensively.

- The other elements for optimization of the value added chain – such as *reducing the number of suppliers (supplier concetration)*, applying the *no-buffer principle* within the establishment *(Kanban)* and introducing *process synchronous (JIT) supply and delivery to the customer* in order to reduce unnecessary storage of materials and buffer supplies – are each employed by between 20 and 30 percent of the establishments. However, in only approximately 5 percent of the enterprises in the investment goods industry are they used comprehensively.

2.4   Diffusion and Structural Characteristics of Manufacturers

There are correlations between the diffusion of new production concepts and the structural characteristics of manufacturers (figure 2-1). Three characteristics were examined: establishment size, type of production, and industry branch.

For *establishment size*, it is more likely that teamwork, continuous improvement (CIP), and the segmentation of production are implemented in large manufacturers. This is most obvious in the conversion of the value-added chain from the performance production arrangement to an arrangement oriented along product lines. Teamwork is less common in small and medium-sized manufacturers.

With regard to the *type of production*, it is clear that the continuous improvement process is more widespread among manufacturers producing to inventory or without customization than in manufacturers with customized production. This could be because the production process in customized production is not sufficiently standardized so that the instrument of CIP can only be used insufficiently. In addition, in manufacturers that produce customized products, the tendency prevails to resolve problems by communication, which could make an introduction of the CIP superfluous. However, it is surprising that teamwork and manufacturing segmentation is more widespread among manufacturers producing according to programs than in manufacturers with customized production. This seems inconsistent to the extent that these concepts aim at ensuring a fast and flexible adaptation of production to customer specific requirements. For manufacturers with customized production there, therefore, still seems to be conversion and improvement potentials available in order to increase speed and flexibility when handling customized orders by introducing the corresponding elements of new production concepts.

Figure 2-1   Diffusion and the structural characteristics of investment goods manufacturers

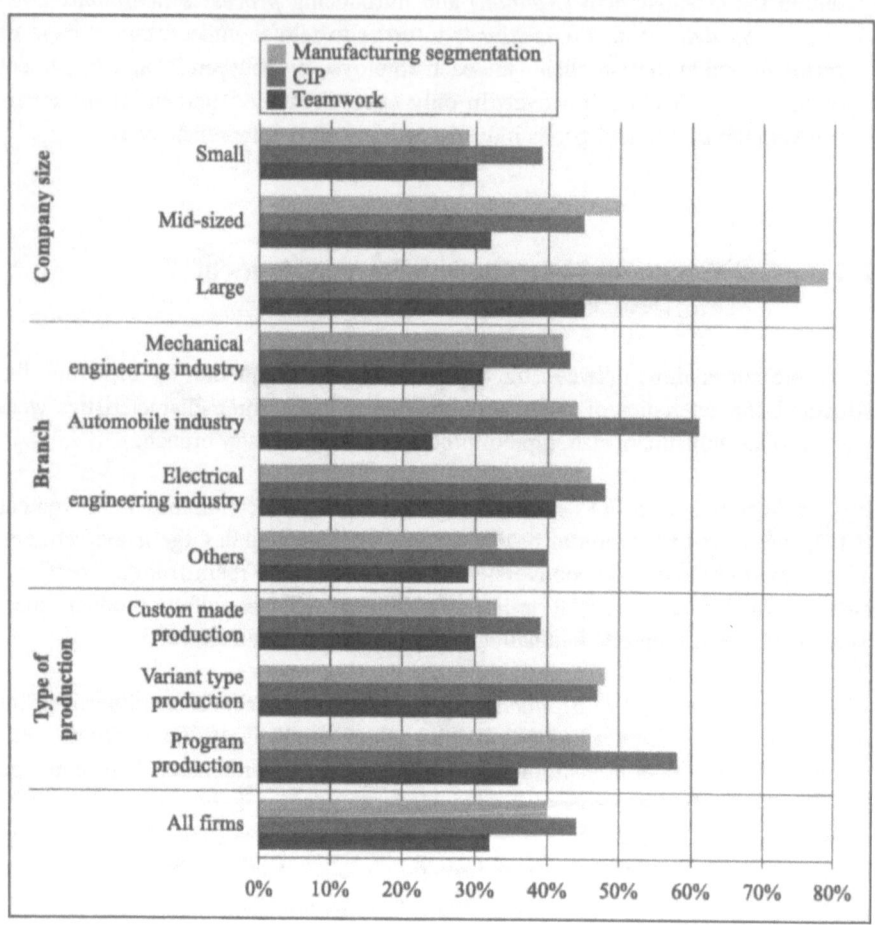

There are also some surprises concerning the *industry branches*. The frequently stated leading role of the automotive industry and its suppliers is only demonstrated in the diffusion of innovative concepts of quality management, such as in the process of continuous improvement (CIP). In the diffusion of organizational concepts such as teamwork, the automotive industry shows the lowest user rates, the electrical industry takes over the leading role here. The degree of diffusion of manufacturing segmentation is almost the same in mechanical engineering, automotive production, and the electrical industry. Manufacturers from other branches of the investment goods sector show lower shares in the diffusion of this element as some of these manufacturers carry out construction-site production and thus, segmentation is out of the question.

2.5 The Dynamics of Diffusion

As our look at the structural features of German investment goods manufacturers shows, not every firm organizes its production according to the new principles. There are product and production features which make the introduction of individual elements of new production concepts impossible or at least unreasonable (Dreher et al. 1995). We have tried to distinguish establishments who cannot use an organization principle from those who could use it but currently do not (figure 2-2). Even taking into account manufacturers who say a principle is not applicable, these results show that there is still much potential for further implementation, particularly in the area of work organization.

Figure 2-2   Diffusion of new production concepts – current and potential use

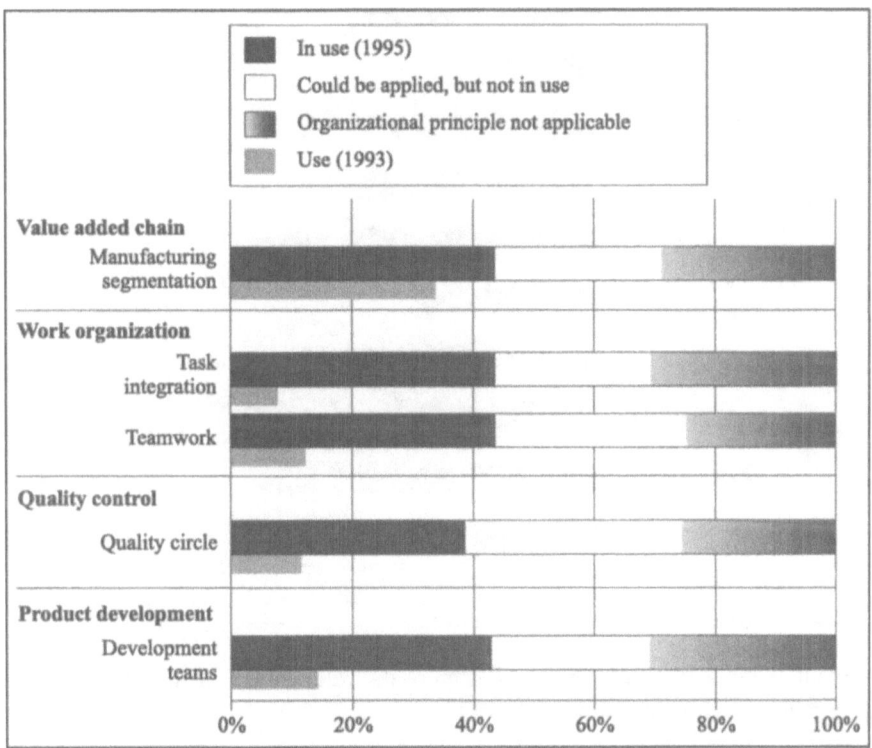

We compared the current data with diffusion data for selected of new production concepts from an earlier ISI survey (figure 2-2). We found a relatively strong increase in the implementation of new forms of work organization, quality control, and product development between 1993 and 1995. The number of manufacturers employing these principles significantly exceeds the levels attained in 1993.

Figure 2-3   Current and planned applications of individual elements of new
                  production concepts

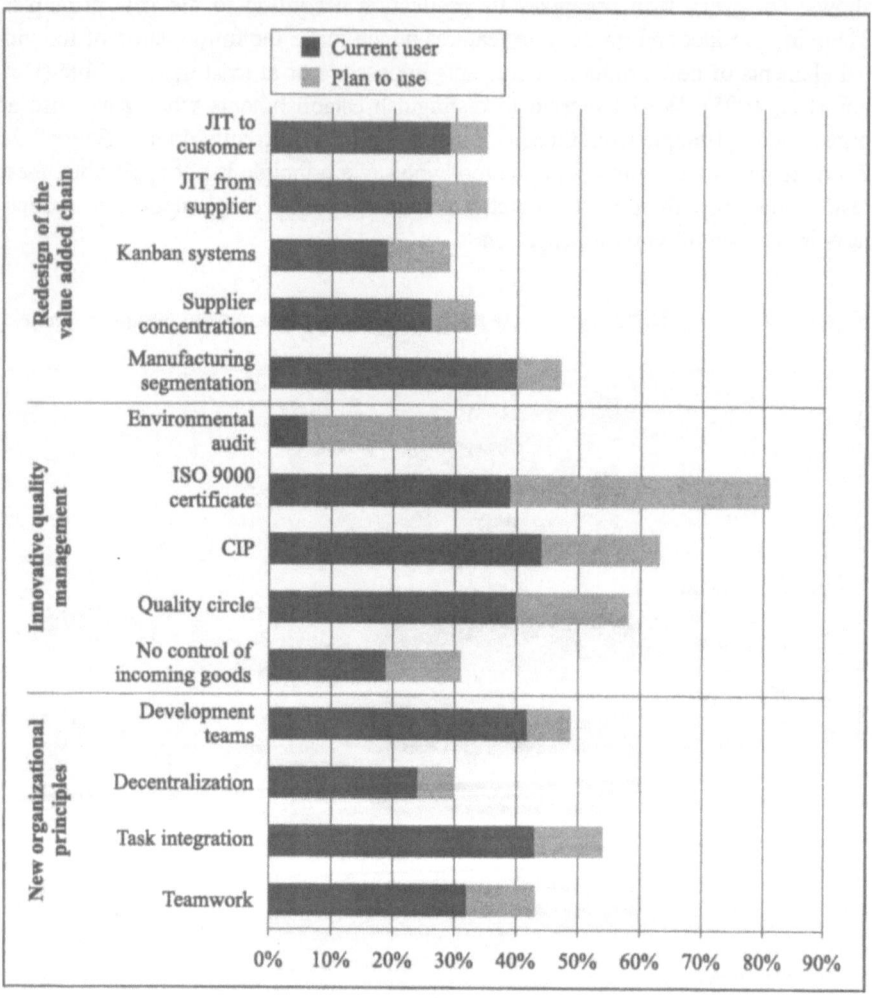

Growth is less noticeable for the redesign of the value-added chain, where the level
of use only slightly increased since 1993.

The plans of the manufacturers, which can be inferred from figure 2-3, show that
high rates of growth are most predominant in the field of quality management
methods. The proportion of users adopting these elements for the first time appears
to be between 15 and 20 percent. As can be observed in the past the measures to
reorganize the value added chain, in contrast, show the least future dynamic. Less
than 10 percent of the manufacturers are even planning to implement the corresponding concepts. Saturation with regard to further diffusion can also be seen in

the organizational design elements. Only one tenth of those asked are still intending to introduce teamwork or an integration of different tasks for the first time. The number of manufacturers planning to apply the other organizational elements of the new production concepts is even smaller. Thus, only a slight increase in the diffusion of these elements can be expected. The dynamics observed at the beginning of the 1990s already appear to have ended.

The greatest proportion of manufacturers intending to be active in the future want to secure a certificate *in compliance with* the ISO 9000 standards. The pressure exerted by customers to obtain such certificates and audits seems to prompt almost all non-users to make plans for the corresponding measures. One fourth of all manufacturers have concrete plans for an audit in accordance with environmental conservation principles. This corresponds to four times the number of manufacturers already using this element. Future growth trends here could be comparable to that observed in the area of with quality.

2.6   Conclusions

Despite the high level of attention to new production concepts recent years and to their potentials in fostering an increase in efficiency and competitiveness, the application possibilities in the investment goods sector have not yet been exhausted. Relatively few manufacturers make comprehensive use of the possibilities offered by the new approaches and principles. The largest unexploited potentials are still to be found in the areas of work organization and personnel management (Kinkel and Wengel 1997, Kleinschmidt and Pekruhl 1994, SFB 187 1991-1995). In addition, after the initial euphoria, the dynamic of their further diffusion seems to have come to a standstill. A one-sided debate, oriented on cost reduction threatens to force the approaches of a comprehensive modernization and innovative orientation of production structures into the background (see Chapter 9).

The different intensities of use of the individual elements of new production concepts depending on the manufacturers basic conditions show the necessity for determining the right mixture and extent of realization according to each individual manufacturer. Besides the branch, the manufacturer size, and the type of production, there are also other factors decisive for the extent of the impacts of new production concepts. To guide the actions of individual manufacturers, the effects of the various elements of new production concepts have to be made transparent against the background of the numerous underlying factors that have to be addressed simultaneously. Similarly, to assess overall progress in the sector, attention must be paid to controlling basic conditions such as product, size of manufacturer, type of production and batch size produced. Of course, not all the elements of the new concepts can be productively realized in every firm (Dreher et al. 1995).

However, even taking this into account, it must be said that if actions speak louder than words, there has been a great deal of discussion about new production concepts, but rather less action.

2.7   Bibliography

Dreher, C., Fleig, J., Harnischfeger, M., and Klimmer, M. (1995). *Neue Produktionskonzepte in der deutschen Industrie*, Physica-Verlag, Heidelberg.

Goldman, S., Preiss, K. and Nagel, R. (1995), *Agile Competitors and Virtual Organization: Strategies for Enriching the Customers*, Van Nostrand Reinhold, New York.

Hammer, M. and Champy, J. (1993), *Reengineering the Corporation*, Harper Collins, New York.

Kinkel, S. and Wengel, J. (1997), Neue Produktionskonzepte: Eine Diskussion macht noch keinen Sommer. Mitteilungen aus der Produktionsinnovationserhebung, PI-Mitteilungen, No. 4, Fraunhofer Institute for Systems and Innovation Research, Karlsruhe.

Kleinschmidt, M. and Pekruhl, U. (1994), Kooperation, Partizipation und Autonomie. Gruppenarbeit in deutschen Betrieben. In: *Arbeit*, 2, 3, pp. 150-172.

Lay, G. and Mies, C. (Eds.) (1997), *Erfolgreich reorganisieren. Unternehmenskonzepte aus der Praxis*, Springer, Berlin, Heidelberg 1997.

Lay, G., Dreher, C., and Kinkel, S. (1996), Neue Produktionskonzepte leisten einen Beitrag zur Sicherung des Standortes Deutschland. Mitteilungen aus der Produktionsinnovationserhebung, PI-Mitteilungen, No. 1, Fraunhofer Institute for Systems and Innovation Research, Karlsruhe.

SFB (various years), Sonderforschungsbereich 187 der Ruhr-Universität Bochum: Mitteilungen für den Maschinenbau, Numbers 1 (August 1991) to 11 (September 1995).

Warnecke, H.J. (Ed.) (1992), *Die Fraktale Fabrik - Revolution der Unternehmenskultur*, Springer, Berlin, Heidelberg, and New York.

Womack, J., Jones, D., and Roos, D. (1989), *The Machine that Changed the World*, Rawson Associates, New York and Toronto.

3   Performance Impacts of New Production Concepts

Gunter Lay, Carsten Dreher and Steffen Kinkel

3.1   Introduction

In the mid-1990s, the economic situation of the German investment goods sector was problematic. Macroeconomic indicators showed that incoming order levels were weak, while major concerns had arisen over the sector's international competitiveness and the loss of industrial employment. The self-assessments of investment goods firms confirmed this discouraging picture. In the Fraunhofer ISI's Manufacturing Innovation Survey (see appendix), one fifth of manufacturing respondents stated that they would not be able to continue to survive if their profitability remained at the present level. One third of the establishments in the survey considered their present profit level as being only sufficient to keep the business alive. Just two fifths of the respondents stated that their present profit level was satisfactory (figure 3-1).

Without long-term improvements in competitiveness and profits, further business collapses are likely in the German investment goods sector, with an associated loss of jobs. To avoid this, and to create incentives for investments that can generate new jobs in this industrial sector, two sets of parallel measures are called for. First, measures to enhance productivity, so that final production costs in Germany are not substantially higher than those in other high wage countries. Second, measures to strengthen non-cost-related competitive factors, such as quality, timeliness, and rapid processing and delivery, to the point where they compensate for any disadvantages in costs.

One strategy that could potentially help to achieve these aims is the introduction of new production concepts (see Dreher et al. 1995). These concepts have been promoted over the last few years under rubrics such as lean production (Womack, Jones and Roos 1989), business reengineering (Hammer and Champy 1993) and the fractal factory (Warnecke 1992). In the previous chapter, Kinkel and Wengel outlined the main elements that comprise new production concepts and examined the

Figure 3-1   Self-assessment of profitability by establishments in the German
                   investment goods sector, 1995

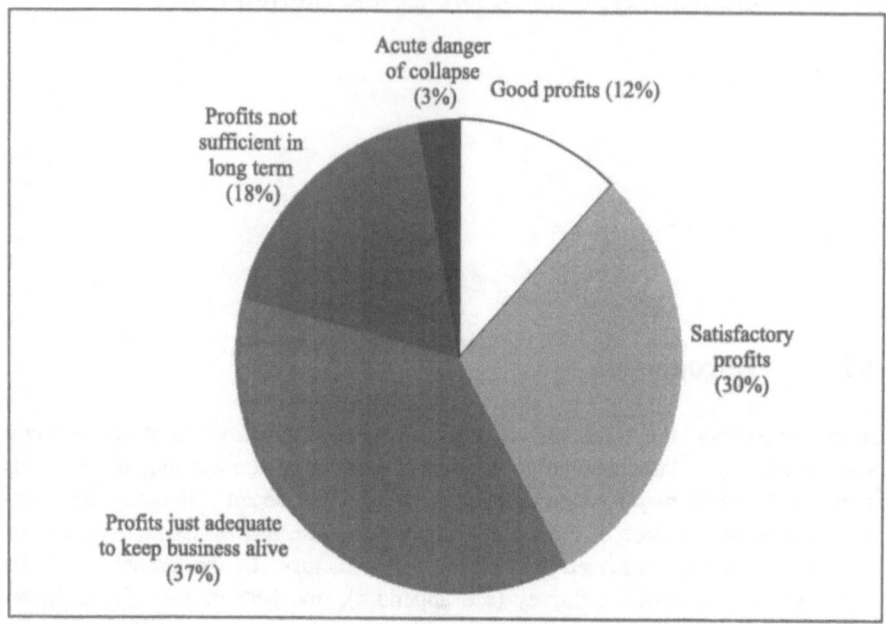

diffusion levels of these elements in the German investment goods sector (see
chapter 2).

In the present chapter, we ask what contribution do these new production concepts
actually make in enhancing the competitiveness of enterprises? Can firms that have
implemented these new principles of organization be differentiated, in terms of
measurable performance indicators, from firms that have not so far adopted them?
These impact questions form the focus for this chapter. We examine the performance impacts associated with the use of new production concepts, using these indicators: productivity benefits, quality improvements, inventory savings and improvements in flexibility. There is also a discussion of how the use of new production concepts affects manufacturers of different size groups, industry branches, and
production approaches. Overall, we find that manufacturers that have introduced
new production concepts are more productive and perform better than non-users.

3.2   New Production Concepts and Productivity

In an industrial context, productivity is the efficiency with which enterprises are able to transform purchased inputs into finished components and products. While the use of modern machinery is an essential element in attaining high productivity rates, in a global business environment where machinery is ubiquitous, further improvements in productivity are increasingly associated with "working smarter" – for example, through enhancements in organizational structures, design for manufacturing, work processes, training, and teamwork.

Productivity is commonly measured in terms of value added per employee, where the numerator (value added) is the difference between sales revenue and the cost of purchased material and service inputs. For the manufacturers responding to the ISI survey, value added per employee averages DM 126,000 (equivalent to about US$ 87,900 at exchange rates prevailing in 1995). However, there is a significant difference in productivity between the group of manufacturers that have implemented elements of new production concepts and those stating that they have not yet done so (Lay, Dreher and Kinkel 1996). These differences are illustrated in figure 3-2.

The greatest differences in productivity are seen for manufacturers using Kanban systems to minimize material buffers between processing stages and for manufacturers using just-in-time methods with their suppliers. In both cases, average value added per employee for users is about 19 percent higher than for non-users of these techniques. Productivity in manufacturers who have adopted decentralized decision making for operational planning is 17.5 percent higher than in manufacturers who have retained centralized operational decision structures. Meanwhile, productivity in plants with quality circles is 11 percent higher than in those without this form of organization.

Although smaller, significant productivity advantages to users are evident for other new production concepts. Value added per employee in manufacturers who have abandoned the highly specialized separation of work tasks in favor of integrated work responsibilities is 9.5 percent higher. By the same measure, productivity is 8.5 percent higher in manufacturers that have introduced teamwork than in manufacturers who have retained traditional forms of individual work allocation. Manufacturers that have changed their manufacturing method from a performance production arrangement to product line-oriented manufacturing segments are 7 percent more productive than manufacturers who have not.

In situations where manufacturers have adopted several complementary elements of new production concepts at the same time, the productivity effects are even greater. For instance, in cases in which teamwork has been combined with the integration of

Figure 3-2   Value added per employee and use of new production concepts
(comparison of average values)

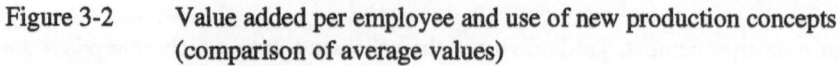

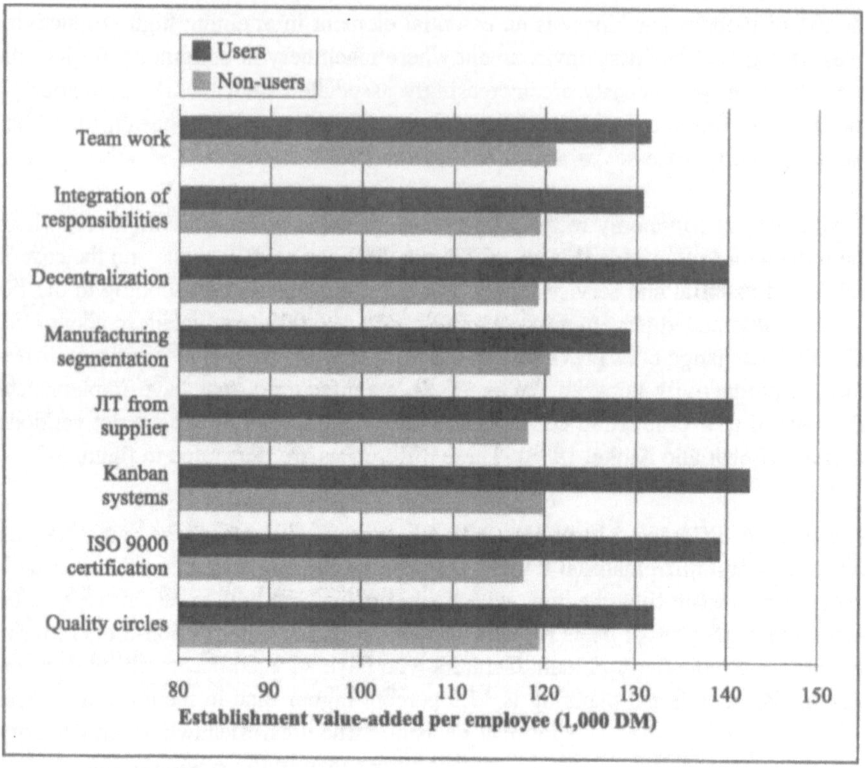

Establishment value-added per employee (1,000 DM)

work responsibilities and with the decentralization of shop floor planning within
manufacturing segments, the value added per employee is approximately DM
154,000. Manufacturers who have not introduced any of these elements average a
value added per employee of DM 117,000. Thus, the productivity lead is over 30
percent.

Although multiple factors affect productivity, in our view the implementation of
new production concepts is one of the major reasons why users of these concepts
have higher productivity levels than nonusers. Follow-up qualitative surveys show
that in the great majority of cases in which fundamental restructuring took place
within a company, this was triggered by a productivity crisis. These manufacturers
recognized that their productivity levels were below par and adopted new production concepts as a key component of their restructuring strategies. Increases in productivity took place after these measures were introduced (Fleig and Mies 1996,
Lay and Mies 1997).

3.3   New Production Concepts and Quality

Customers generally look for both keen pricing and high quality in the products they purchase. Thus, while high productivity is a critical contributor to competitiveness, successful companies also have to ensure that they maintain, if not improve, the quality standards of their products. So, how do new production concepts affect quality?

One criterion of the quality standard of a production system is the proportion of pieces that are rejected or reprocessed. Overall, the average reject rate for all manufacturers in the ISI survey was 4.9 percent. Three new production concepts are most likely to affect quality: quality circles, continuous improvement, and quality certification (ISO 9000). We would expect users of these three elements to demonstrate lower than average quality reject rates than nonusers. The results from the ISI survey show that this is the case (see figure 3-3).

Figure 3-3   Reject rates (percent) and use of new production concepts (comparison of average values)

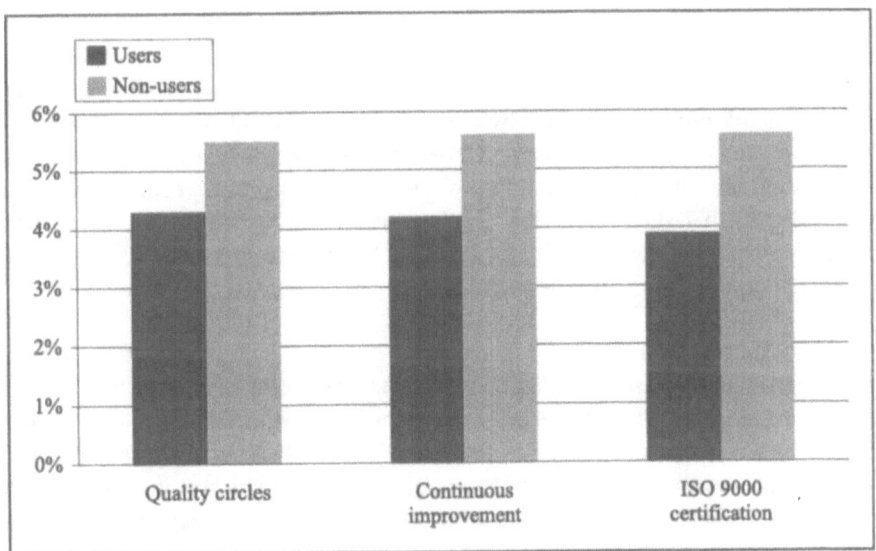

Measured by the reject rate, we find that the use of quality circles results in an average reduction of 1.2 percentage points compared with nonusers. The introduction of continuous improvement processes cuts down the rate of rejects by 1.4 percentage points compared with plants not using this method. Finally, the activities necessary for ISO 9000 quality certification resulted in an average reduction of 1.7 percentage points in the overall rate of rejects compared with non-certified plants.

Overall, these differences represent an intrinsic improvement of 30 to 40 percent in the rate of rejects.

3.4 New Production Concepts and Inventory Savings

A large inventory ties up capital, thus creating costs. It thus makes good business sense to reduce inventory without, of course, generating bottlenecks in production or affecting the ability to deliver. This presents a problem of optimization. In the past, firms often maintained high in-process and final inventory levels, in order to ensure the continuity of production and timely delivery. Today, in the search for higher efficiency and competitiveness, firms seek ways to reduce inventory and thus reduce carrying costs. In addition, inventory reductions often help to highlight previously masked production deficiencies or problems in quality and delivery that, if resolved, can further boost performance.

Three new production concepts are most oriented to reducing in-process and final inventories. These are just-in-time delivery from suppliers, the use of "kanban" principles, and just-in-time delivery to customers. Under just-in-time principles, supplies and orders are delivered close to or at the point they are actually required in a manufacturing process or by a customer. This customization requires tighter scheduling of orders, manufacturing, and logistics, but also reduces expensive warehouse stocks and lowers waste. Under kanban principles, manual or electronic systems are used to ensure that a step in the production process occurs only when demanded by the successive step. Again, this avoids the expensive build-up of work-in-process on the factory floor and final warehouse inventory, as well as of-fering opportunities to reduce waste.

We have already seen (section 3.3) that such new production concepts are associ-ated with reduced wastage through lower reject rates. But how do these concepts affect inventory levels? On average, the manufacturers in the ISI survey carried inventory levels that would suffice for 41 days of production. However, manufac-turers who used the new production concepts of kanban and just-in-time relations with their suppliers and customers were able to manage with significantly fewer inventories (figure 3-4).

When kanban is implemented, the level of inventory (measured by how many days of production that the inventory could support) is reduced by eight days on average, which is in turn a decrease of about 20 percent. Just-in-time delivery from suppliers reduces inventory stocks by more than 20 percent on the average when compared with nonusers. Meanwhile, manufacturers that use just-in-time when delivering to customers have lower inventory levels of 15 percent.

Figure 3-4   Inventory levels (in days) and use of new production concepts
(comparison of average values)

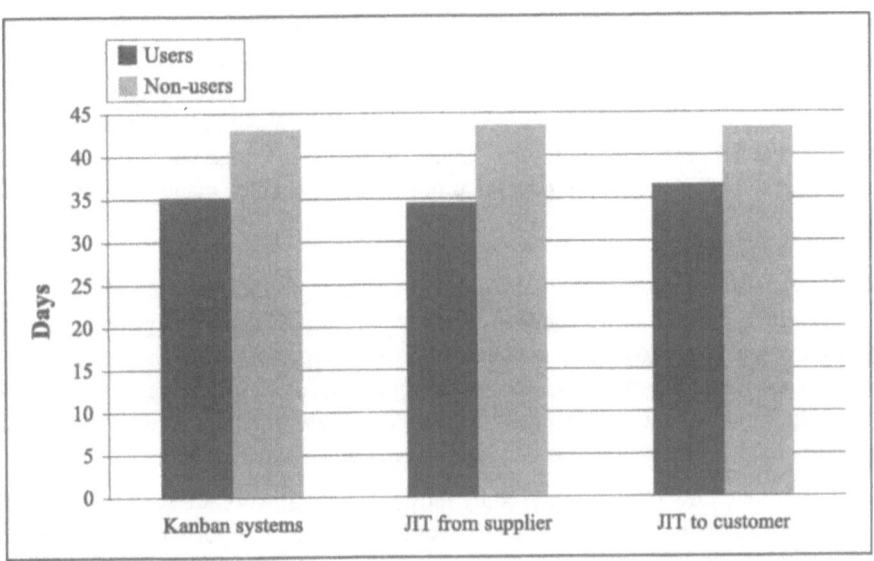

The firms using these new production concepts will benefit from the reduced capital costs that result from lower inventory levels. Some of these concepts also reduce waste, as noted earlier. However, there are trade-offs. The absence of material buffers and the introduction of just-in-time delivery are also associated with longer processing times. Individual manufacturers thus have to decide for themselves what their priorities are, and consequently what elements of the new production concepts are most suitable for them.

3.5   Comparative Impacts of New Production Concepts by Industry and Employment Size

So far, we have reported the impacts of new product concepts as averages across all industries and employment size classes within the investment goods sector (after weighting the raw survey responses by the scheme described in the appendix). But it is apparent that there are major differences in productivity levels by industry, employment size, and other factors within the investment goods sector. How do these differences affect the impact of new production concepts?

Underlying variations in the performance characteristics of respondents are quite evident in the ISI survey. The average value added per employee is approximately DM 133,000 for the mechanical engineering manufacturers responding to the survey, DM 138,000 for automotive manufacturers, DM 116,000 for electrical engineering plants. The larger the manufacturer, the higher the average value added per employee: it increases from DM 122,000 for manufacturers with less than 100 employees to DM 150,000 for manufacturers with more than 500 employees. Nonfinancial differences are also apparent. For example, average inventory stocks totaled 40 days in mechanical engineering, 49 days in electrical engineering businesses, and 28 days for automotive plants. In larger manufacturers, the materials stored were sufficient for 28 days of production, in smaller manufacturers for 39 days.

These variations in performance by industry and size do influence the effects associated with the use of new production concepts. Analyses carried out in this respect show that the new production concepts improve the performance and competitiveness of manufacturers in all of the sectors and categories of size surveyed. However, the *extent* of the improvement that can be made in productivity, quality and material buffers depends on industry and size. Here are two examples:

- The productivity effects achieved by introducing teamwork are strongest in large manufacturers (with a workforce of over 500). In this category, the difference in value added per employee between manufacturers with and without teamwork is DM 27,000. In small and medium sized manufacturers, the productivity potentials created through teamwork are definitely lower, since here the unproductive elements of high task specialization (number of interfaces, unused capacities, doubling of tasks) are obviously not as strongly present as in large manufacturers. The introduction of teamwork thus has less potential for change.

- The inventory reduction effects achieved by just-in-time supply were most significant in the automotive industry and in mechanical engineering. In these industries, the difference in the inventory stored by manufacturers employing just-in-time supply and those not amounted to nine days of production. The differences in other sectors were not as noticeable.

These two examples demonstrate that, when comparing industry and firm sizes, the improvement potentials of the various elements of the new production concepts are not always highest where productivity and performance indicators are lowest.

3.6 Importance of Specific Elements of New Production Concepts for Individual Manufacturers

Besides industry and employment size, there are other factors that determine the impact of new production concepts. To provide a reliable basis for action in an individual manufacturer, the likely effects of various elements of new production concepts must be made transparent. This means considering as many aspects as possible about a manufacturer's initial situation, as shown by the two following examples, which analyze the effects of teamwork in two types of mechanical engineering businesses. The two types differ by product, employment size, type of operations planning and production batch sizes.

- Plant Type 1: Medium-sized mechanical engineering establishments (100 to 500 employees) which manufacture complex individual products according to specifications of customers.

- Plant Type 2: Small mechanical engineering establishments (less than 100 employees) which manufacture relatively simple products in medium-sized series; the products are varied from time to time according to specific customer requirements, but do not have to be totally redesigned for each order.

We compared the *productivity effects* of teamwork in these two types of manufacturers (figure 3-5). We found that in Type 2 manufacturers the improvement impacts of teamwork are definitely higher, namely 23 percent, than in Type 1, in which the difference found between manufacturers employing teamwork and those not was 12 percent. The reason for this difference may be that in Type 2 plants, products are less complex and production processes are more easily rationalized, meaning that work teams can more readily achieve improvements. In contrast, in Type 1, due to complex products and customer-specific manufacturing, productivity depends on know-how and expensive specific tools, meaning that the scope for productivity improvements under the control of a work team is smaller.

In a comparison of the *effects* of teamwork *on the quality* for the two types of manufacturers, clear quality advantages from teamwork again can be seen in Type 2, whereas in Type 1 significant quality effects were not found. With quality measured by the rate of rejects, there is no difference between users and nonusers of teamwork in Type 1. However, in Type 2 the rate of rejects is 2 percent when teamwork is not employed and only 0.6 percent when it is employed. The reason for this may be that in traditional specialized mechanical engineering plants (Type 1), quality-conscious skilled workers are mostly employed, and their orientation towards high-quality production cannot be noticeably increased through teamwork. In serial production, on the other hand, teamwork is a way to achieve process-synchronous quality control and allows the employees, who are often semiskilled, to become more knowledgeable about and responsible for quality, resulting in substantial quality improvements (Type 2).

Figure 3-5   Use of teamwork and productivity, quality, and flexibility, for Type
1 and Type 2 manufacturers (average values)

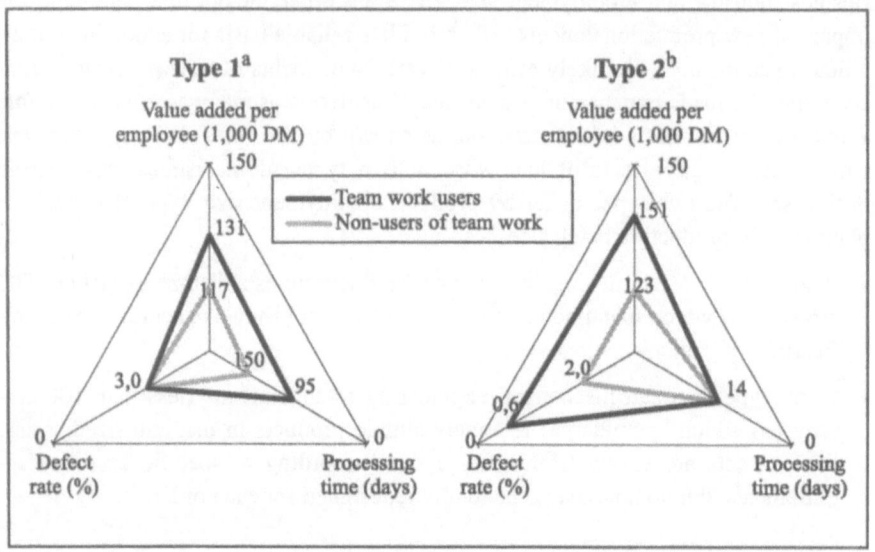

ᵃType 1: Medium-sized mechanical engineering establishments (100 to 500 employees) – manufac-
turing complex individual products – to customer specifications.

ᵇType 2: Small mechanical engineering establishments (less than 100 employees) – manufacturing
simple products, medium-sized series – some product variation, but not redesigned for each order.

We also compared the *flexibility effects* of teamwork between Type 1 and Type 2
manufacturers, using processing time as a measure of flexibility in production. The
following pattern was noted: in Type 1, the overall processing time is determined
by the date of order and by the time required to obtain necessary parts, as well as by
the handling time. The introduction of teamwork mainly affects handling time
within the firm to such an extent as to reduce processing time by more than 30 percent. The reduction of interfaces associated with the introduction of teamwork is
particularly profitable in medium-sized manufacturers who have to carry out detailed planning for complex products. In Type 2, these aspects are much less important. The manufacturing of simple articles requires less preparation; efficient
pre-production is not a problem when controlling a production program that only
has to take limited customer-specific variations into consideration. Thus it is not
surprising that employing teamwork in this type of firm does not have a great impact on processing time.

market strategies and situations within manufacturers have an influence on the success of new production concepts. Moreover, not all performance indicators can be improved to the same extent and at the same time. Thus new production concepts do not represent a total solution for all industrial problems.

3.7   Conclusions

From the evidence presented above, it is clear that manufacturers who have already introduced elements of new production concepts are more efficient than other manufacturers in terms of productivity and performance. This holds true for a very wide variety of industries in the investment goods sector and is also independent of the firm size. Thus, the introduction of new production concepts should be regarded as a serious alternative by manufacturers whose critical income situation makes the relocation of their production to low-wage countries appear as the only remaining alternative. (For a further discussion of this point, see chapter 8).

The introduction of individual elements of the new production concepts also needs a preliminary individual analysis of the importance of specific competition factors regarding the firm concerned, since the effects of these elements can definitely be ambivalent, depending on the situation in the firm. Thus, it is essential to be acquainted with the firm's competitive strategy. With this knowledge, an individual analysis using the data foundation described here can be designed using the experiences of comparable manufacturers in such a way as to achieve a direct guideline to be followed for the individual firm.

3.8   Bibliography

Dreher, C., Fleig, J., Harnischfeger, M., and Klimmer, M. (1995), *Neue Produktionskonzepte in der deutschen Industrie: Bestandsaufnahme, Analyse und wirtschaftspolitische Implikationen*, Physica-Verlag, Heidelberg.

Fleig, J. and Mies, C. (1996), Multidisziplin - Erfolgreiche Unternehmen in einem turbulenten Umfeld, *Die Mitbestimmung*, 6, pp. 47-50.

Hammer, M. and Champy, J. (1993), *Reengineering the Corporation*, Harper Collins, New York.

Lay, G. and Mies, C., (Eds.) (1997), *Erfolgreich reorganisieren. Unternehmenskonzepte aus der Praxis*, Springer, Berlin and Heidelberg.

Lay, G., Dreher, C., and Kinkel, S. (1996), Neue Produktionskonzepte leisten einen Beitrag zur Sicherung des Standortes Deutschland. Mitteilungen aus der Produktionsinnovationserhebung, PI-Mitteilungen, No. 1, Fraunhofer Institute for Systems and Innovation Research, Karlsruhe.

Warnecke, H.J., (Ed.) (1992), *Die Fraktale Fabrik – Revolution der Unternehmenskultur*, Springer, Berlin and Heidelberg.

Womack, J., Jones, D., and Roos, D. (1989), *The Machine that Changed the World*, Rawson Associates, New York and Toronto.

4   Interaction of Process and Product Innovation

Gunter Lay

4.1   Introduction

The tapping of new markets with growth possibilities is often said to require the development and commercialization of innovative products (Utterback 1994, Freeman and Soete 1997). A strategy of innovation and technology leadership appears to be the most promising way, especially for manufacturers in high-wage Germany, to not only maintain, but also to expand their sales performance and market positions. In turn, the ability to innovate to ensure continued success in sales and markets is closely linked to the preservation of manufacturing jobs in Germany.

The relationship between product innovation and business growth is clearly illustrated in the results of the ISI manufacturing innovation survey of the German investment goods sector. We compared the sales performance of manufacturers across three groups: establishments that had not introduced product innovations in the previous three years, those with a low share of product innovations (up to 25 percent of sales), and those with a high share of product innovations (more than 25 percent of sales). We found that as manufacturers increased their share of innovative products, they were also more likely to experience growth in sales (see figure 4-1).

Many factors influence a manufacturer's ability to develop and commercialize innovations. Much attention has been focused on the role of research and development within firms, the transfer of new knowledge and technology, links with users, access to finance, industrial and competitive conditions, and the role of strategic management and entrepreneurship (Teece 1986, Dosi et al. 1988, Freeman and Soete 1997. But also material to a manufacturer's ability to innovate is the organization within the enterprise of product development departments and groups, technical facilities, and links to manufacturing operations, as well as the structure of cooperation between product developers and others within and outside of the company. To address these factors, many companies have adopted new organizational

Figure 4-1   Sales growth and the share of innovative products, German
                    investment goods sector, 1995

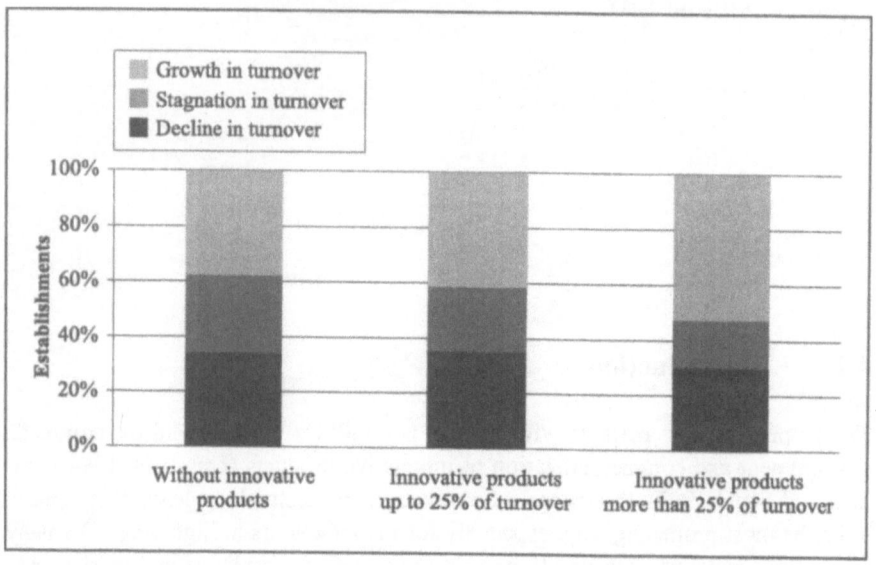

approaches to innovation in the last few years, as part of the broader introduction of
new production concepts.

This chapter examines the extent to which the specific use of new production concepts and associated organizational approaches contributes to innovation performance. The chapter draws on the experience of the German investments goods sector
reported through the ISI Manufacturing Innovation Survey. There is an exploration
of the relationships between the use of new production concepts and the level and
pace of product innovation. The hypothesis is tested that process innovations are a
key to innovative products and so to turnover growth and the safeguarding of jobs.
The chapter also probes the correlation between process and product innovations
independently of establishment characteristics (see also Pleschak and Sabisch
1995).

4.2   New Production Concepts and Product Innovation

When asked in 1995, about two-thirds (62 percent) of manufacturers in Germany's
investment goods producing sector said that in the previous three years they had
developed and introduced innovative (i.e. new to them) products. This is similar to
results found in other studies (see SFB 1997). There were variations in innovative-

ness by size. At small establishments with up to 100 employees, 56 percent had developed and introduced innovative products. This figure rose to 69 percent at medium-sized units (100 to 300 employees) and to 74 percent at larger plants. Differences between industries were also observed. Of the establishments in the electrical engineering industry, 70 percent had introduced product innovations to the market, compared with 68 percent for machinery manufacturers, 53 percent for manufacturers of other metal products, and 53 percent for structural metal product makers.

In addition to these variations by industry and size, equally large differences were apparent if the firms were divided into users and non-users of new production concepts. Four new production concepts that were likely to bear on product innovations were examined:

- *Simultaneous engineering* – which aims to shorten product development times by concurrently implementing steps in the product innovation process that are usually pursued consecutively (see Womack, Jones and Roos 1992, p.121). Over one-quarter (27 percent) of establishments in the investment goods industry use simultaneous engineering, although four percent of these respondents say they are using this concept only on a pilot basis.

- *Interdepartmental development teams* – these overlay or supercede functional intra-departmental project groups with cross-cutting teams comprised of staff from different units within a company, aimed again at accelerating and improving the product innovation process (Womack, Jones, Roos 1992, p.119). Over two-fifths (42 percent) of investment goods industry plants say they have adopted interdepartmental development teams. Of these, four percent are trying this concept on a pilot basis.

- *Cooperation in research and development with suppliers or customers* – so as to better tailor product innovations to customer needs and to fully exploit supplier capacities (see Wolff et al. 1994, p.47). This is the most widely diffused of the four new production concepts analyzed in this section. Almost one half (47 percent) of establishments report that they engage in this form of co-operative development. Of these, seven percent are only pursuing this on a trial basis.

- *Continuous improvement processes (CIP)* – through which company personnel, not only from design departments but also from manufacturing, work together to overcome bottlenecks and enhance the product development process (see Robinson 1991). More than two-fifths (44 percent) of investment goods producers have institutionalized CIP, of which four percent say that this is on a pilot basis.

There is a connection between the use of these new production concepts and the orientation towards innovation. Manufacturers that fully use one of these four production concepts (excluding pilot users) are much more likely than non-users to develop and introduce new products (figure 4-2). Eighty percent of the plants using simultaneous engineering have developed product innovations and introduced them to the market. The corresponding figure in plants not using simultaneous engineering is 57 percent. More than three-quarters (76 percent) of the manufacturers with interdepartmental development teams had undertaken new product development, compared with just over one-half (53 percent) of non-users of this concept. For manufacturers who cooperate in research and development with customers or suppliers, 77 percent had undertaken product innovations, whereas for non-users it was 53 percent. Finally, 73 percent of the plants adopting CIP achieved product innovations. For non-users, the comparable figure was 56 percent.

Figure 4-2   Product innovations and use of new production concepts

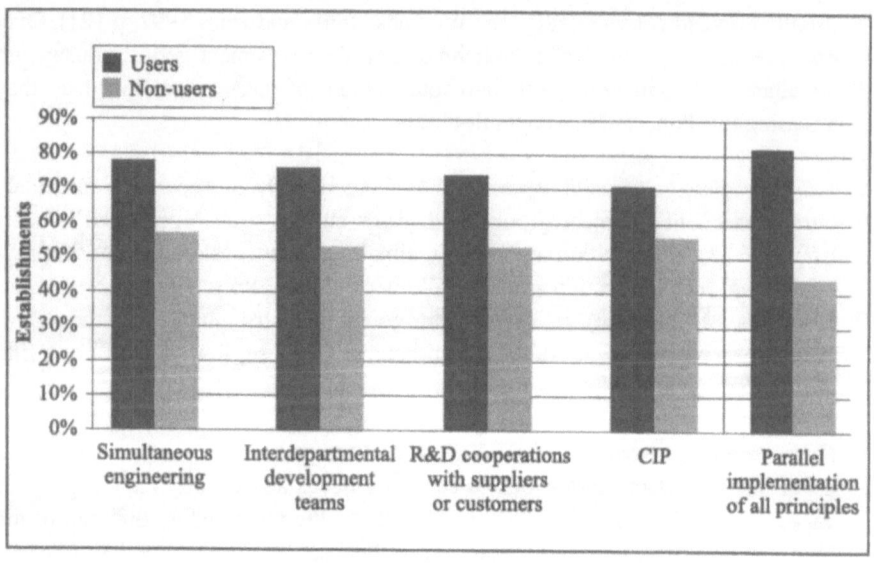

The connection between the adoption of new production concepts and an orientation to product innovation is made even clearer when manufacturers who use all the above mentioned elements are compared with those who use none. In plants where all four new production concepts, 82 percent undertake product innovation; this is nearly double the rate of product innovation than found in plants which use none of the four new production practices (44 percent).

The linkage between innovative processes and products holds true independently of the size or industry of the manufacturer or of the complexity of the product. Only the dimensions of the positive effects associated with process innovation appear to vary, in three notable ways. First, the smaller the firm, the clearer is the lead of process-innovative companies in the development and market introduction of innovative products. Second, the less complex the manufactured product is, the more open are process-innovative firms to product innovation. Third, in firms in electrical engineering and the heterogeneous group of "other" investment goods industries, the positive effects of process innovations on the generation of innovative products is greater than for example in the mechanical engineering industry, although the connection can also be made here as well.

4.3　Influence of New Production Concepts on Product Development Timing

The time that a firm requires developing an innovative product and introducing it to the market is generally regarded as an important determinant of successful innovation leadership. The shorter the time to market, the greater the probability of gaining a competitive advantage.

However, development times for product innovations depend on a number of factors that can only be partly influenced by the firm. If the time needed to develop new products in the investment goods industry is examined, it can be seen that it differs clearly according to the complexity of the manufactured products, the dependence on the customer order for development and the size of the manufacturers (figure 4-3). Manufacturers of less complex products stated that they required 9 months to develop an innovative product. Manufacturers of products of medium complexity reported an average of 13 months development time. Manufacturers of complex products need on average 16 months to bring a product innovation to market maturity.

Manufacturers who start product development on commission after a specific customer order reported that an average of 13 months is necessary for such customer-specific new product development. Plants that develop innovations independently of customer orders, for example for a production program from which the customer can then order variations, require on average 15 months development time for a product innovation. Manufacturers from whom the customers can only choose from a product program, reported an average of 18 months time necessary for a new product development. Finally, smaller plants with up to 100 employees state that an average of 13 months is required for product innovations, medium-sized manufac-

Figure 4-3   Product innovation development times and product complexity, development type, and establishment size

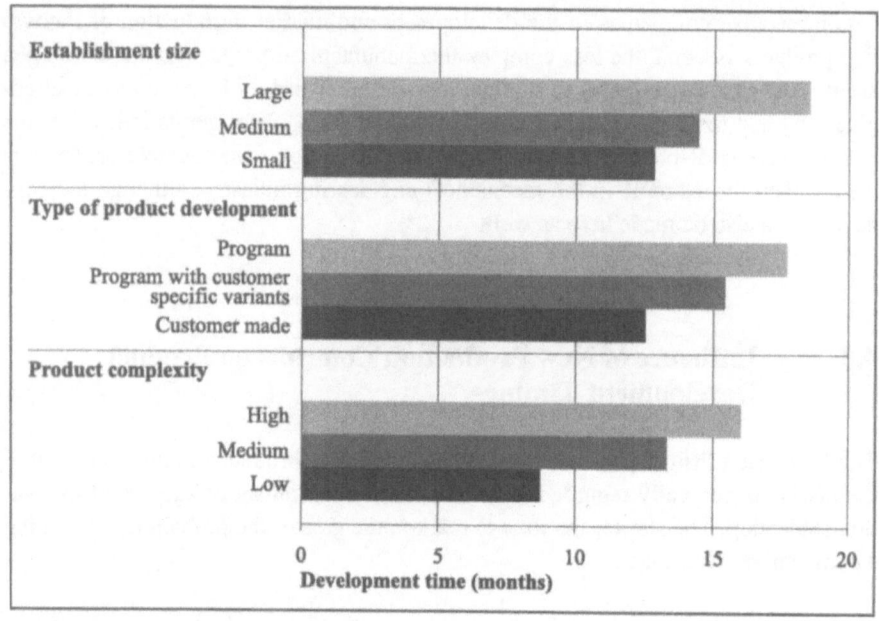

facturers (100 to 300 employees) an average of 14 months and larger manufacturers (over 300 employees) an average of 19 months.

These differences make it necessary, when analyzing the influence of new production concepts on the development times for product innovations, to compare only those groups of firms that are comparable. From the data available in the ISI survey, five categories of firms have been identified, clustered according to size, product complexity, and level of product customization. (To avoid problems associated with small sample sizes, each of these categories contains not only a minimum of 30 ISI survey respondents, but also includes more than 10 establishments who report they are users or non-users of new production concepts.)

- *Type A* – small establishments that have products of medium complexity in their product program and undertake new product development for a production program that is developed without a specific customer orders. The average development time for product innovations in this group is 13 months.

- *Type B* – small establishments that develop and manufacture highly complex products to customer order. New product development in this group requires an average of 13 months.

- *Type C* – small establishments that manufacture highly complex products and develop production programs with customer-specific variations. In this group, the average time required for new product development is 15 months.

- *Type D* – medium-sized establishments that manufacture highly complex products and design product innovations for a production program with customer-specific variations. An average of 18 months is required for new product development in this group.

- *Type E* – larger establishments that manufacture products of great complexity and undertake product innovations as part of a production program with some variations, but not truly custom-made. Establishments of this type need the longest time for new product development, with an average of 21 months.

To what extent does the time taken to bring new products to market for these five types of manufacturers differ according to whether the plant has implemented new production concepts? The new production concepts studied here are ones that are likely to influence the product development process, namely: simultaneous engineering, the use of computer aided design (CAD), and the formation of interdepartmental development teams (see Eigner and Maier 1985, Womack, Jones, Roos 1992). The results from this analysis are illustrated in figure 4-4.

Simultaneous engineering proved, in four of the five establishment types analyzed, to be an organizational concept that shortens the time to market for product innovation. For example in medium and large plants that use simultaneous engineering (as full users), the time to market for new product development is up to 14 percent shorter than for non-users. Only in Type B (small plants with high product complexity and customer-specific development) was there no shortening of the time to market associated with simultaneous engineering. In fact, development time increased by 36 percent, a reversal of the assumed effect.

Computer aided design was also accompanied by shorter times to market in the same four manufacturing types. Typically, a 10 percent reduction in time to market was associated with CAD, although among large plants (Type E), a reduction of 24 percent was reported. There was again a reversal in the expected relationship for Type B. CAD users in this group reported times to market that were 18 percent above those of other plants in the group which did not use this technology.

Inter-departmental development teams were associated with shorter times to market for product development by all five groups of establishments. In plants where interdepartmental development teams had replaced functional organizational arrangements for specific projects, reductions in time to market of about 10 to 15 percent were observed, compared with plants not using this tool. In large plants, the time saving (up to 40 percent) was even greater.

Figure 4-4   Development times for new products, by use of selected new
                production concepts.

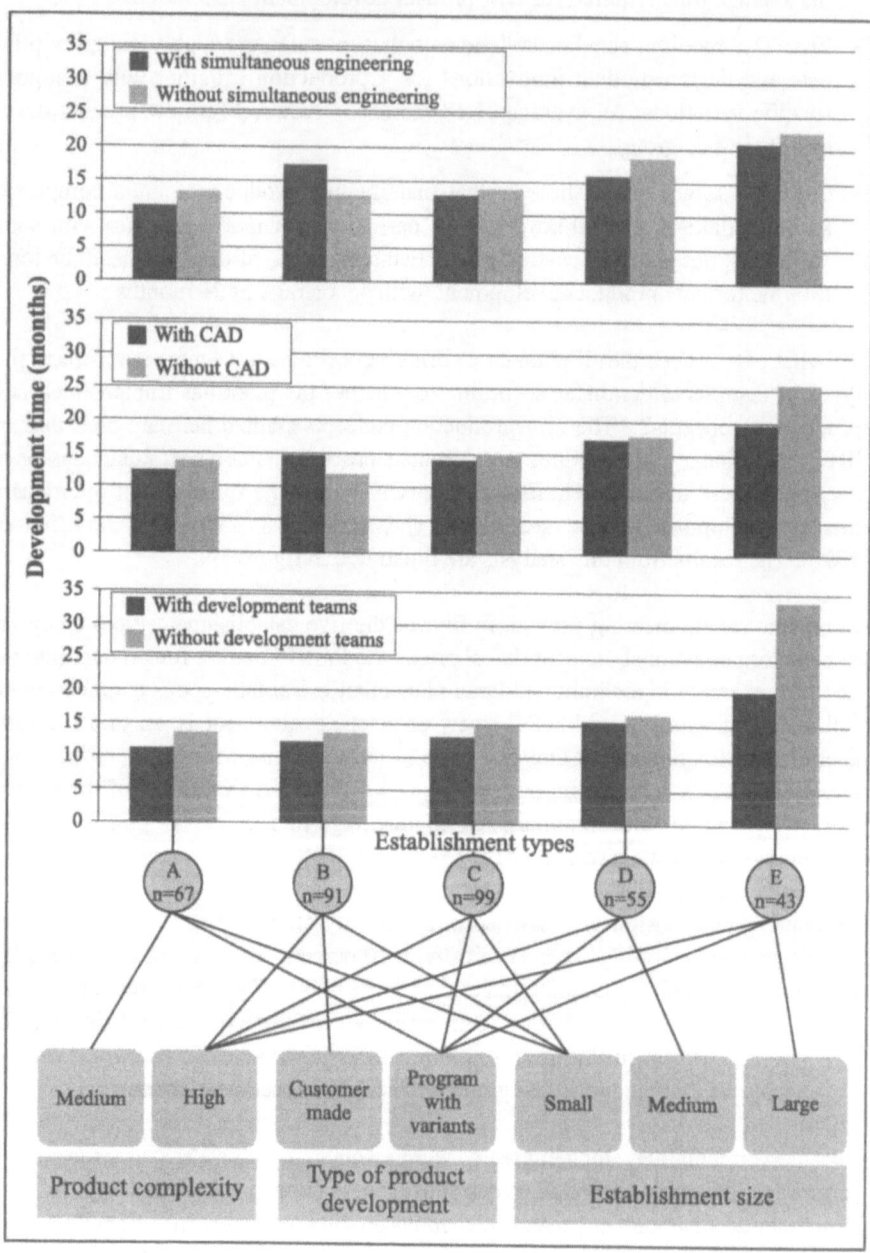

In general, it seems that the use of new organizational production concepts is associated with shortening in new product development time to market among manufacturers with different characteristics and product development needs. The potential for time saving is greatest in medium-sized and large plants making complex products. Moreover, although the results for Type B seem to be an exceptional reversal, on closer inspection there are wide differences in the time to market reported within this class by those new product developers who also use new organizational production concepts. These wide differences are not present among non-users of new production concepts in class B, or among plants in the other four groups. These statistical differences need to be kept in mind when interpreting the group B results.

4.4   Conclusions

The analysis that is reported in this chapter shows that the performance and pace of new product development in the investment goods sector is related to the use of new organizational production concepts. For most types of manufacturers, the use of new organizational production concepts like simultaneous engineering, CAD, or inter-departmental work teams, speeds up the product development process. The impacts on product development gained from using new organization production concepts are greatest in those companies that are presently furthest behind in their innovative performance. Generally, small manufacturers lag larger ones in product innovation. However, small manufacturers that employ new production concepts are more likely to perform product innovation than other small manufacturers who do not use these new production concepts. Small manufacturers using new production concepts are thus helped in narrowing the gap in innovation performance with larger plants.

What do these findings mean for business strategy? Of course, new production concepts can be used to rationalize production processes to offset the labor cost disadvantages of high-wage production in Germany. However, we have also seen that new production concepts can be used to improve the process of new product development. With the help of the new production concepts, German companies can compensate for the high cost of domestic manufacturing by emphasizing competitive factors other than the product price (see also Lay 1997). Thus, an advantage may be maintained by offering a range of highly innovative products and by developing those products in a short time.

If companies gear their competitive business strategy towards innovation and technology leadership, then the results presented here show a clear, positive connection between the targeted use of coordinated elements of new production concepts and innovation performance. In particular, there is shortening in the time required to bring new products to market – an increasingly important factor in meeting customer needs and stimulating new demand. Process innovations thus become a key to innovative products. New production concepts can help enterprises to improve their innovative ability and to venture into market sectors that are not dominated by cost competition alone and which can provide new possibilities for growth.

4.5   Bibliography

Dosi, G., Freeman, C., Nelson, R., Silverberg G., and Soete, L. (Eds.) (1988), *Technical Change and Economic Theory*, Pinter, London.

Eigner, M. and Maier, H. (1985), *Einstieg in CAD*, Carl Hanser Verlag, München.

Freeman, C., and Soete, L. (1997), *The Economics of Industrial Innovation*, MIT Press, Cambridge, MA.

Lay, G. (1997), Wer kein Ziel hat, verzettelt sich. In: Lay, G. and Mies, C. (Eds.) *Erfolgreich Reorganisieren*, Springer Verlag, Berlin, pp. 43-68.

Pleschak, F. and Sabisch, H. (1996), *Innovationsmanagement*, Schäffer-Pöschel Verlag, Stuttgart.

Robinson, A. (1991), *Continuous Improvement in Operations*, Productivity Press, Cambridge, MA.

SFB 187, (1997), NIFA-Panel der Ruhr-Universität Bochum, Mitteilungen für den Maschinenbau, No. 17, Bochum, September.

Teece, D. (1986), Profiting from technological innovation: implications of integration, collaboration, licensing and public policy, *Research Policy*, 15, 6, pp. 285-305.

Utterback, J. (1994), *Mastering the Dynamics of Innovation*, Harvard Business School Press, Boston, MA.

Wolff, H., Becher, G., Delpho, H., Kuhlmann, S., Kuntze, U., and Stock, J. (1994), *FuE-Kooperationen von Kleinen und Mittleren Unternehmen*, Physica-Verlag, Heidelberg 1994.

Womack, J., Jones, D., and Roos, D. (1992), Die zweite Revolution in der Autoindustrie, 7. Auflage, Frankfurt.

5   New Production Concepts and Employment

Gunter Lay

5.1   Introduction

With more than 4 million people out of work and an unemployment rate of over 11 percent in Germany, the creation of new jobs is now a major concern. One approach that is often suggested to lower unemployment is the use of new production concepts in industry (European Commission 1997). The aim is to improve the competitiveness of manufacturers, and thus their sales, through such methods as employee teamwork, decentralized decision-making, orientating production towards customer requirements, and combining previously fragmented tasks. It is hoped that the increase in turnover generated by these methods will enable new employees to be hired, thus helping to reduce joblessness (Blechinger et al. 1997, p.74).

Of course, encouraging manufacturers to transition to new ways of organizing production is desirable for reasons other than those of job creation. But from the view of employment policy, does the use of new production concepts actually result in new jobs? Also, manufacturers may pursue new production concepts with different aims in mind; for example, some manufacturers want to reduce costs while others seek to improve their performance in terms of better quality, innovative capability and faster, more flexible delivery times. Thus, there might there be differential effects on employment, depending on the strategic orientation of enterprises and how they implement new production concepts.

Since many German manufacturers have now implemented new production concepts, employment changes in these establishments can now be compared with what has happened in plants that have not so far introduced new forms of production organization. A basis for comparison is provided by the 1995 ISI manufacturing innovation survey of the German investment good sector. The review of the employment effects of new production concepts contained in this chapter uses this database to answer the following questions:

- In general, does employment increase in manufacturers that use new production concepts when compared with nonusers or low-users of these concepts?

- When exploiting the potential of the new production concepts, are alternative strategic business orientations – namely, pursuing a cost reduction strategy versus a performance or quality improvement strategy - associated with different employment effects?

- How does a cost-oriented strategy differ from a performance-oriented strategy in the implementation of particular new production concept elements?

- What is the quantitative importance – including effects on employment - of the different ways in which new production concepts are being integrated into business strategies in Germany today?

5.2 Development of Employment in Manufacturers: Users and Nonusers of New Production Concepts

A series of new production concepts were probed by ISI in its 1995 survey of the investment goods industry in Germany. These included simultaneous engineering, temporary interdepartmental development teams, R&D co-operation with suppliers and customers, continuous improvement processes (CIP), production segmentation, "no buffer" principles (just-in-time) in internal material flows, the decentralization of planning, management and control functions, employee task integration and employee teamwork. (See the appendix to this book for details about the survey.)

In the following analysis, establishments are classified according to their self-reported data on how many of the specified elements they have introduced so far. If an establishment's numerical use of new production concepts is below the median, it is classified as a nonuser/low user. If deployment is at or above the median, the plant is classified as a user of new production concepts.

Since the size of a manufacturer affects the extent to which elements of the new production concepts are implemented, specific median values have been used for each establishment size group. Smaller units (with fewer than 50 employees) need only to implement more than two elements of new production concepts in order to be classified as users. In large plants (with 1,000 or more employees), seven or more of the listed elements have to be present for the plant to be classified as a user. This procedure ensures that the user and nonuser/low user categories are similarly distributed by establishment size. Thus, different employment trends between these categories cannot be the result of size-specific differences. The results of the analysis are shown in figure 5-1.

Figure 5-1   Employment trends among users and nonusers of new
              production concepts

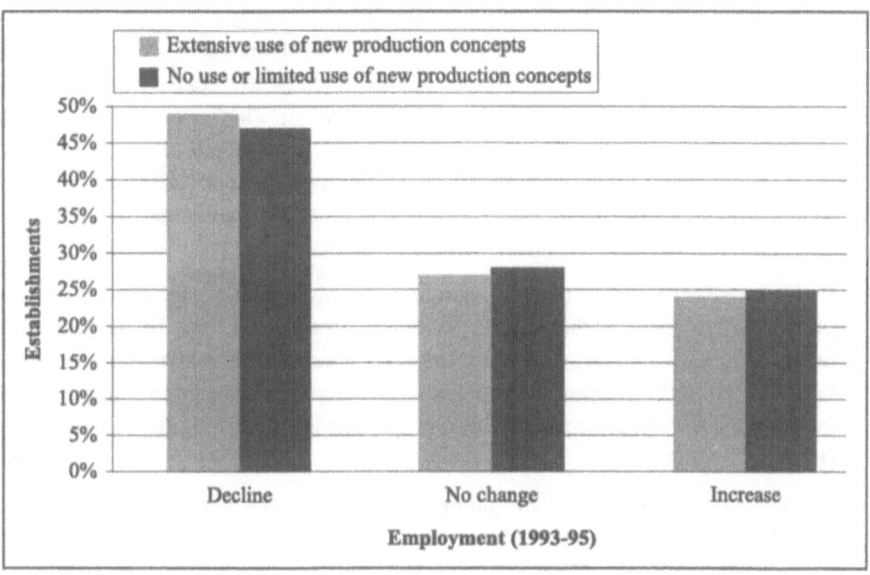

As these results show, the survey evidence does not confirm the broad hypothesis that introducing new production concepts leads to positive employment impacts. There are only marginal differences in employment change between manufacturers operating with or without new production concepts. Admittedly, the survey examined a short time period over which users of new production concepts had relatively limited experience on which to report. During this period, it seems that the job rationalization effects of new production concepts were not compensated by additional turnover to the extent that additional employment was created.

5.3   Employment, New Production Concepts, and the Effects of Alternative Business Strategies

In further analysis, the relationship between the use of new production concepts and job creation was examined in the context of whether the manufacturer was pursuing a cost-based or a performance-based strategy. Additionally, an assessment was made of the effects of each strategy on the manufacturer's competitive position. It was found that employment effects are largely dependent on the type of business strategy being pursued by the manufacturer (see Brödner et al. 1997). Four strategic case scenarios (A through D) were hypothesized that combined different permutations of business strategy and market success with impacts on employment. These scenarios were then examined empirically using the survey results.

Assume that if a manufacturer uses new production concepts to lower production costs in order to maintain price competitiveness, it can then reduce its workforce by releasing employees who are no longer required due to the productivity improvements generated by the new production concepts. In the short-term, wage costs go down and products can be offered at a lower price. In the medium term, employment development in manufacturers where new production concepts have been applied in a strategy of this kind may follow one of two alternative courses:

• If the manufacturer succeeds in reducing its wage costs so far that the lower-priced product gains additional market share, turnover will rise. If the additional turnover is high enough, the extra employees needed to realize this increased turnover will then compensate or even over-compensate for the initial job losses incurred by the deployment of the new production concepts in a cost-oriented strategy (Type A).

• If lower prices do not gain the manufacturer additional market share and turn-over growth, the introduction of new production concepts in this type of strategy will lead to job losses at the level of the manufacturer. But the remaining jobs, while fewer, may be more secure, since the manufacturer is at least in a position to stabilize its turnover through improved price competitiveness instead of suffering an absolute turnover decline (Type B).

In the second set of case scenarios, it is assumed that new production concepts are exploited not in a cost-oriented strategy but in order to offer customers better quality, shorter or more reliable delivery times, innovative products or more customized solutions. This orientation will probably mean that the "saving potentials" are lower. Personnel will become superfluous – although less than in the cost-oriented strategy. However, presume that these employees are not be made redundant but are re-deployed in order to improve the manufacturer's performance in the non-cost-oriented competition factors mentioned above. Here, too, it is the position that the manufacturer can achieve relative to its competitors that will ultimately decide the employment effects within the manufacturer, with two variations:

• If the manufacturer is merely enabled by this strategy to keep up with its competitors, it will stabilize its turnover and its workforce will stay the same (Type C).

• If the manufacturer is able to gain competitive advantages through a performance improvement strategy, it will attract additional market share, its turnover will increase and it will need additions to its workforce to cope with this growth in turnover. Employment will thus rise (Type D).

As these scenarios indicate, the use of new production concepts in manufacturers may result in differential job impacts, ranging from decreased employment (Type B), constant employment (Types A and C), or new job creation (Type A and D). The actual impact on employment is thus decided by the manufacturer's strategic orientation in implementing the new production concepts and the market position that subsequently results.

Using data from the ISI questionnaire, the following section examines how far these four theoretical cases of the alternative employment impacts of the introduction of new production concepts correspond with reality. Responding establishments are clustered to correspond to the four cases suggested above, allowing a comparison of employment effects to be made (see figure 5-2).

Figure 5-2   Employment trends, new production concepts, and strategic
                  orientation

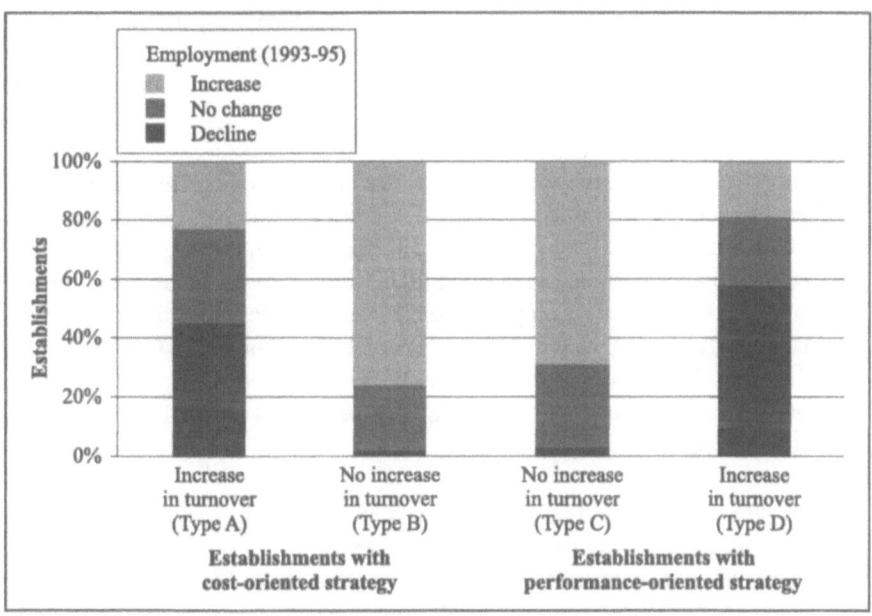

The first group (Type A) comprises manufacturers that introduced elements of the new production concepts primarily with the aim of cost reduction and have succeeded in achieving turnover growth. In this group of manufacturers, 45 percent have experienced employment growth, 32 percent maintained the size of their workforce, and 23 percent report that their employment has declined.

The second group (Type B) comprises manufacturers that have introduced new production concepts with the primary strategic aim of lowering production costs but have experienced stagnation or even a decrease in turnover because they have failed to increase their market shares. For this group of manufacturers, only 3 percent have taken on additional employees, while 22 percent have nevertheless maintained their workforce. However, 76 percent of manufacturers in this group have reduced employment.

In the third group (Type C) were manufacturers that had introduced new production concepts not with the overriding aim of reducing production costs, but with a strategic orientation towards performance factors such as quality, flexibility or short delivery times. In this group of manufacturers, production lead times were 25 percent shorter and reject rates 10 percent lower than the comparable values for manufacturers following a cost-oriented strategy. The stronger orientation towards customer requirements was expressed in significantly smaller batch sizes. However, manufacturers in group C were able only to maintain their competitive position, with turnover remaining at the same level or, in some cases, experiencing a decline. Of the manufacturers in this group, only three percent reported an increase in employment, 29 maintained their prior employment levels, and 69 percent reduced employment.

As in the C case, the fourth group of enterprises (Type D) comprised manufacturers in which the introduction of new production concepts was part of a performance-oriented differentiation strategy. However, in Type D manufacturers, a better competitive position has been achieved through this strategy, with increases in turnover being attained. Their production performance characteristics were similar to Type C manufacturers, but there were differences in employment outcomes. Of the group D manufacturers, 58 percent reported increased employment and 23 percent said their employment had remained the same. Only 19 percent of plants in this group reduced their workforce.

These results demonstrate that practice tends to confirm the theoretical considerations described above regarding the strategic dependency of the impacts on employment when new production concepts are implemented. Manufacturers that improve their performance with new production concepts and increase their market share experience more positive employment outcomes than manufacturers that successfully use new production concepts within a cost-oriented strategy. On the other hand, manufacturers that are unable to improve their market position through their selected strategy in using new forms of production – or that even fail to maintain their position – are, as one would expect, not in a situation to offer additional jobs. However, where failure (to increase market share) occurs, employment reductions are significantly larger in manufacturers that pursue cost-oriented strategies than in manufacturers with performance-oriented strategies.

From an employment policy viewpoint, it is therefore preferable if enterprises located in Germany attempt to strengthen their performance with the new production concepts rather than use these concepts to reduce their costs. While a successful cost-oriented strategy is better in terms of employment effects than an unsuccessful performance-oriented strategy, the best employment outcomes are found through the successful implementation of new production concepts with a performance-oriented business strategy.

5.4 Differences between Cost-Oriented and Performance-Oriented Business Strategies in Shaping Production Concepts

When considering the deployment of new production concepts in manufacturers and strategically related employment outcomes, attention needs to be paid to the specific ways in which individual new production concepts are implemented. Three dimensions can be distinguished:

First, individual *elements* of new production concepts may be *differently organized* depending on strategic context within the manufacturer. This means that very different organizational models may be concealed behind labels such as "teamwork" or "temporary interdepartmental development teams." Thus, the implementation of teamwork in a cost-oriented strategy may not be associated with consistent qualifications and skills for all team members, enabling each member to fulfil all tasks occurring within the group. Rather, the group may be structured to make the team as small as possible. By contrast, teamwork in a performance-oriented strategy implies the deliberate creation of quantitative and qualitative "redundancies" (i.e. more than minimal levels) in order to ensure that quality and the ability to deliver remain unaffected, even in potential bottleneck situations.

Second, the *selection and combination* of elements from the "tool box" of new production concepts may vary depending on a manufacturer's strategic orientation. In a cost-oriented strategy, emphasis may be given to elements such as the no-buffer principle in the internal flow of materials, the institutionalization of continuous improvement processes and task integration to cut down on personnel. By contrast, a performance-oriented implementation of new production concepts might focus production linkages with product innovations, simultaneous engineering, new co-operative partnerships with customers and suppliers, and the formation of temporary task-related development teams.

Third, differences in strategic orientation can mean that the *savings* from deploying new production concepts may be *put to different uses*. For instance, in a cost-oriented strategy, savings in personnel achieved through decentralization and the dismantling of hierarchical structures may be used to reduce the workforce as a whole. Alternatively, in a performance-oriented strategy, these same measures with similar potential savings may be used to re-deploy excess middle managers into new, customer-oriented activities within the manufacturer.

Thus, it may be possible to distinguish strategic uses of new production concepts that tend to create jobs from those that may cost jobs, insofar as data on these aspects can be gathered in a written questionnaire. Drawing on the ISI survey results, the following points emerge:

- If the introduction of new production concepts with primarily cost-related aims is compared with an implementation strategy oriented towards improving the performance of the manufacturer, there are only slight differences in the specific elements of new production concepts used in each case.

- Those differences that are present, while they are not large, point in the following direction: performance-oriented implementations of new production concepts tend to place more emphasis on teamwork, decentralization, production segmentation and simultaneous engineering. By contrast, in cost-oriented strategies, the no-buffer principle in material flows within the manufacturer is used with above-average frequency (see figure 5-3).

- When probing the hidden significance of the label "teamwork" in a cost-oriented or performance-oriented introduction of new production concepts, small differences are also found. These differences appear to uphold the hypothesis briefly outlined above. In 62 percent of the cases where manufacturers pursue performance-oriented strategies while introducing new production concepts, "teamwork" means that planning and quality assurance tasks are integrated into the work spectrum of the group and that all team members are similarly qualified to perform all group tasks. By comparison, this definition applies in 56 percent of cases where a cost-oriented strategy is involved.

5.5   Importance of Cost-Oriented and Performance-Oriented Strategies in the Realization of New Production Concepts

Bearing in mind the results described so far, what has been the predominant strategic orientation within which new production concepts have been introduced into the German investment goods industry? The following picture emerges (figure 5-4).

Figure 5-3   Strategic orientation and the use of elements of new production
concepts

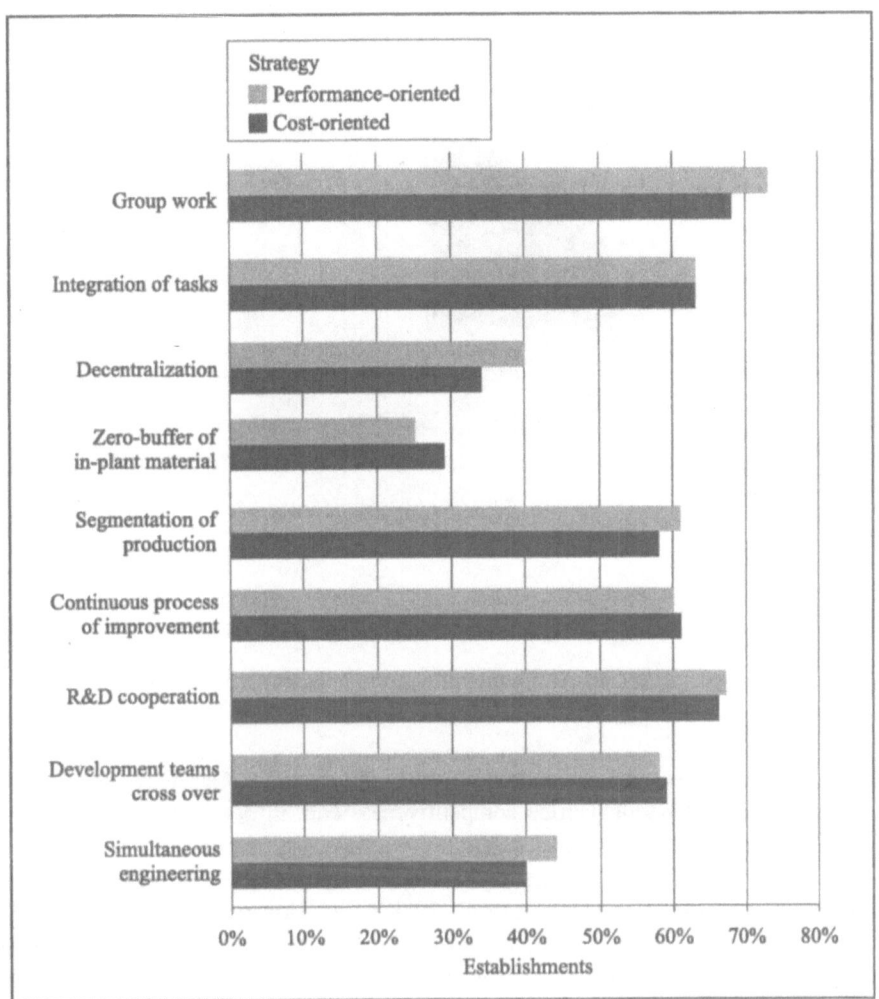

- 57 percent of investment goods manufacturers deployed new production concepts with a strategy oriented primarily towards reducing costs. Of these, approximately two-fifths succeeded in improving their market position and thus
enlarged their turnover. Three-fifths were unable to increase their turnover.

- 43 percent of the users of new production concepts introduced these organization
principles with strategies oriented towards improving their performance. In this
group, somewhat more than half the manufacturers succeeded in increasing their

Figure 5-4   Quantitative importance of cost-oriented and performance-oriented strategies in the implementation of new production concepts

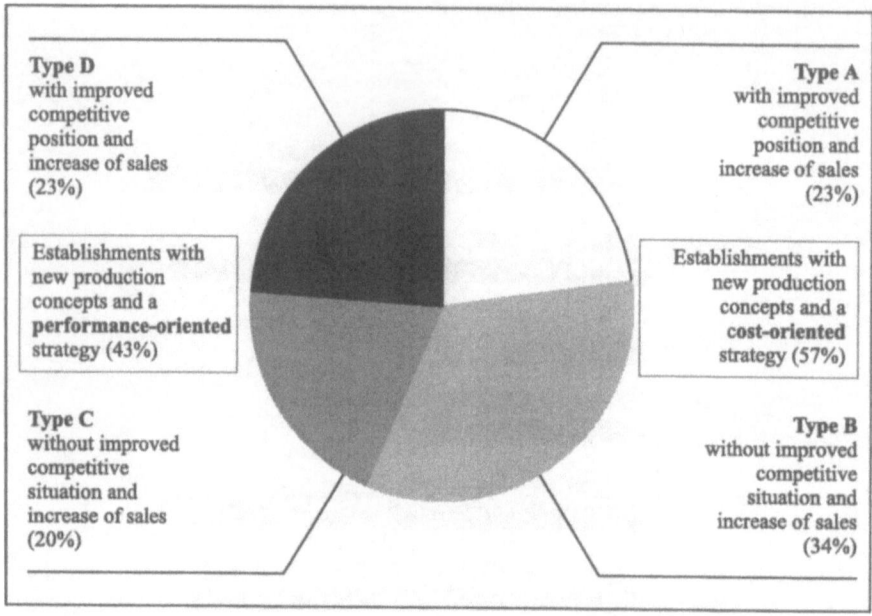

turnover. Just under half of the manufacturers adopting this strategy failed to increase their market share.

New production concepts are more frequently introduced in Germany with the aim of improving international price competitiveness and, in particular, meeting competition from low-wage countries. (For comparison with earlier surveys that have investigated the strategic orientation of manufacturers when introducing technical process innovations, see Matzner et al. 1988 p.40.) However, our results suggest that this goal can only be partially realized, since many manufacturers deploying new production concepts within this strategic orientation are not successful in their attempts to become cost leaders and thus fail to increase their turnover. Thus, new production concepts do not give the impetus to employment that is anticipated.

5.6   Conclusions

This chapter makes it clear that care needs to be taken in assessing the contribution of new production concepts to job creation in Germany. The simple hypothesis that "the increased use of new production concepts will lead to more employment" is not confirmed by empirical data contained in ISI's survey of investment goods producing manufacturers, at least for the time period examined in the survey. Instead, a more disaggregated formulation is necessary – one that takes into account how differing business strategies and conditions affect the relationships between new production concepts and employment.

Thus, the factors that determine whether the introduction of new production concepts in manufacturers results in positive or negative employment outcomes are the strategic orientation with which the concepts are realized and the competitive position that the manufacturer is able to achieve with its strategy. Positive impacts on employment are more often found in manufacturers that have implemented the new production concepts in a strategy oriented towards improving performance and in this way have succeeded in increasing their share of the market. Moreover, the likelihood of opening up additional markets is greater with a performance-oriented strategy than with a cost-oriented strategy.

At present, new production concepts tend to be regarded by manufacturers as instruments of rationalization to deal with cost competition. Yet, this is counterproductive from the viewpoint both of employment policy and business success. We need to consider what incentives can be provided to encourage manufacturers to place higher priority on using new production concepts as part of a performance-based strategic business orientation (see Lehner et al. 1998 p. 479).

5.7   Bibliography

Blechinger, D., Kleinknecht, A., Licht, G., and Pfeiffer, F. (1997), *The Impact of Innovation on Employment in Europe, European Commission*, Directorate General XIII, EIMS Publication No. 46, Brussels.

Brödner, P., Garibaldo, F., Oehlke, P., Pekruhl, U., and Rouilleant, H. (1997), *Work Organization and Employment - The Crucial Role of Innovation Strategies*, unpublished discussion paper (second version), IAT, Gelsenkirchen, October.

European Commission (1997), *Partnership for a New Organization of Work*, Green Paper, European Commission, Directorate General V, Brussels.

Lehner, F., Baethge, M., Kühl, J., and Stille, F. (1998), Beschäftigung durch Innovation - Perspektiven und Ansätze für eine strukturelle Erneuerung von Wirtschaft und Arbeit in Deutschland. In: Lehner, F., Baethge, M., Kühl, J., and Stille, F. (Eds.), *Beschäftigung durch Innovation*, Rainer Hampp Verlag, München und Mering, pp. 463-492

Matzner, E., Schettkat, R., and Wagner, M. (1988), *Beschäftigungsrisiko Innovation? Arbeitsmarktwirkungen moderner Technologien*, Edition Sigma, Berlin.

6   Worker Participation and Process Innovations

Jürgen Wengel and Werner Wallmeier

6.1   Introduction

The involvement of workers in corporate supervision and management – through procedures for "codetermination" – is a well known, and much discussed, feature of the German industrial system. The German model of management-worker relationships is institutionalized in the "Betriebsverfassungsgesetz" (business constitution law) and the "Mitbestimmungsgesetze" (laws for codetermination). In addition to these legal requirements, employee participation is also viewed as an instrument of management. The direct participation of workers in management decisions aims to promote increased acceptance by employees of organizational and technological changes and to mobilize the knowledge and abilities of employees in the interests of the company (Ritter 1992).

In recent years, there has been increased interest in Germany in going beyond conventional forms of co-determination (for example, through employee-management boards) to involving workers in operational decisions in more decentralized and flexible ways. The reason for this lies in the changed internal and external conditions in manufacturing today. Through increasing internal complexity and dynamics, hierarchical leadership-styles have reached their limits. Decentralized coordination procedures (for example, through participation oriented leadership-styles or team approaches) seem more suitable than top-down hierarchical management (see Staudt 1996 and Schanz 1992). In addition, through new methods of work organization and the technical possibilities of computer-aided production, narrow and divided approaches to structuring work are losing relevance. To gain the full benefit from new work methods and technologies, it is important for companies to enable workers to make decisions and assume responsibility and to tap the reservoir of experience and practical knowledge of their employees. Finally, the meaning of work is changing. Employees hope for and expect more possibilities of participation and responsibility in their jobs (Ritter 1992).

The positive effects claimed for direct employee participation include the increased contentment and motivation of workers, the increased acceptance of decisions resulting in less resistance, and a stronger identification of employees with the company. It is also suggested that – compared to other forms of decision making – improved decisions result when these are mutually made by management and workers. Undoubtedly, the identification of employee teamwork as a factor of success of the Japanese production model has contributed to the growing interest in new participatory organizational forms. Interestingly, in the 1970s, the advocacy of factory-level participatory models that emphasized the humanization of work led to little real change in German industry. But, in the 1990s, German firms are now more actively interested in employee participation in their factories, with the aim that this will help to improve their business performance and international competitiveness.

This chapter examines the diffusion of participatory approaches in German industry and explores whether the postulated impacts of participation can be shown empirically. As in other chapters, the analysis is based on the Fraunhofer ISI manufacturing innovation survey of establishments in the investment goods sector (see the appendix for details of the survey). The chapter pursues two principal themes. First, whether manufacturers that make more use of participation differ from those without any participatory orientation with respect to business performance. This analysis is based on a broad understanding of participation that is best described as a "participation-oriented organizational culture". Second, whether direct participation by employees in team decision making processes improves the implementation and returns from other techno-organizational changes within the factory.

6.2 Participatory Organizational Culture and Business Performance

Outsiders who only look at what is contained in the German legal framework can easily overestimate the diffusion of codetermination and employee participation in German industry. Particularly in small and medium sized enterprises, which represent the majority of companies and jobs, the workers' council is not always elected, although the Betriebsverfassungsgesetz gives employees in companies with more than twenty employees this right. Similarly, the financial participation of employees (through employee share ownership) is limited to relatively few individual companies or to large companies in which the employees receive non-voting staff shares. This is despite legislation to encourage employee financial participation ("Vermögensbildungsgesetz") and recent discussions about increasing the "investment wage" in Germany.

However, as the ISI survey shows, there is a growth in strengthening workers' interests in business success though the use of performance-related pay. About one-half of all manufacturers in the survey use performance-related remuneration. This reaches beyond traditional piecework systems. For example, 11 percent of companies say that they use operating profits as one of the criteria in their performance-linked remuneration systems in production.

Additionally, as they adopt new production concepts, many companies have pursued organizational solutions that rely on stronger participation of employees in operational decision processes, management task, and the decentralized delegation of responsibility. As might be expected, however, there are differences in how various types of participatory approaches are deployed:

- Approximately one third of the investment goods manufacturers have adopted quality circles. However, only a little more than ten percent of these respondents use this approach throughout their facilities.

- Almost 40 percent of the manufacturers use principles of continuous improvement processes (CIP). Of these, about 15 percent use CIP comprehensively and not only in parts of the factory or in pilot projects.

- Integrated work groups, with rotating tasks, multi-skills, and integrated planning and quality control are found in one third of the manufacturers. Groups are usually given the responsibility to elect their own leader.

- Codetermination over the distribution of working time is still in an experimental stage at only a few pilot manufacturers. Less than five percent of manufacturers have gathered first experience with teams that have autonomy over their working time.

As in the diffusion of other innovative technologies, larger manufacturers lead in the implementation of participatory forms of organization. For many small and medium-sized establishments, the implementation of formal participatory organizational structures is not judged essential, as these manufacturers often have high levels of informal employee involvement.

What is the difference in performance between, on the one hand, participation oriented manufacturers which have simultaneously adopted various participation forms in parallel and, on the other hand, non-participation oriented manufacturers which have completely refrained from any of the above listed forms of involvement and participation of employees? Figure 6-1 shows the average performance indicators of the two different groups:

- The productivity in participation oriented establishments is almost 20 percent higher on the average.

- The proportion of rejected and reworked parts amounts to an average of four percent compared to six percent in manufacturers without any participation.

Figure 6-1   Participatory orientation of organizational culture and business performance

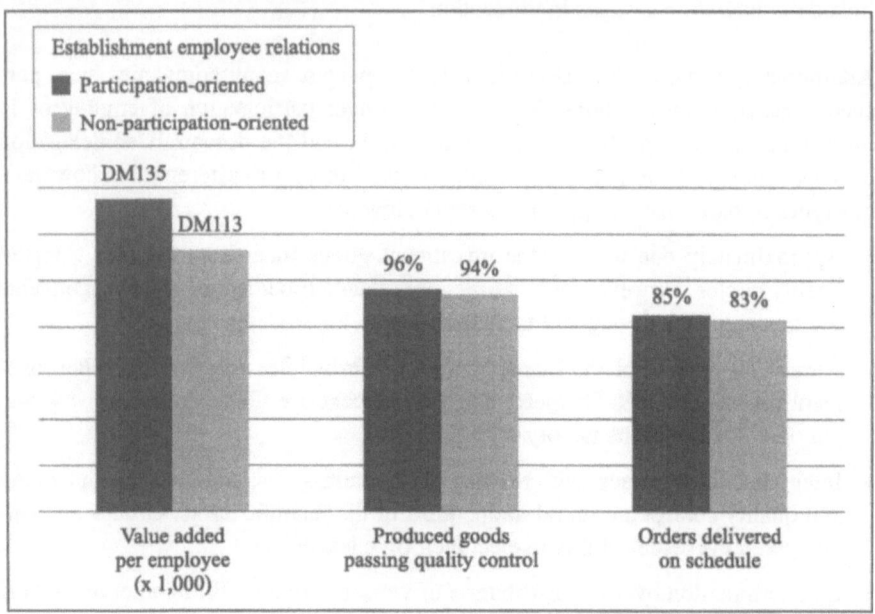

- The level of adherence to deadlines is two percent higher in participation oriented manufacturers.

These differences can also be seen when analyzing subgroups of manufacturers of different sizes or manufacturing structures that usually show different business performance. For instance, large manufacturers usually have a higher net value per employee on the average and more frequently adopt participation-oriented organizational solutions. However, the higher productivity of manufacturers with a participation oriented organizational culture is not based on this correlation. It is to be found in all employment sizes classes.

Nevertheless, these results should not be directly interpreted in such a way as to suggest a simple cause-and-effect-relationship between participation and the performance of German investment goods manufacturers. It is just as likely that more successful manufacturers use employee participation to a greater extent. Moreover, viewed another way, it is clear that involving employees in decision-making does not impose costs or other disadvantages that make participatory manufacturers poorer performers than non-participatory ones (despite occasional media reports that suggest "too much" employee participation is a drain on competitiveness).

It would, however, be premature to expect that in each individual case the solution for decision and leadership problems lies in participation. Greater participation is only one approach to decision making and motivating employees. Participation, thus, competes as a means of improving motivation with financial incentives, flexible working-time regulations, or prospects of promotion. When considering improvements in operational decisions, there are different forms of participation as well as contrasts between participation and other organization forms.

For example, one could consider limited consultation of employees after which the decision remains with management alone, or the complete delegation of decisions to one employee or to a group of employees. Hierarchical decisions also have certain advantages compared to participation oriented decision-making. In the case of the hierarchical decision structures, only instructions have to be passed on to the employees and not the entire information, on which the instruction is based, thus expenses for information are lower (see Casson 1994). Under such short-term considerations, participatory decision-making is not necessarily better than nonparticipatory decision-making. Therefore, the question is not: Does participation of employees pay off or not? The result will depend on the particular case considered. If one assumes that the positive effects of employee participation only appear under certain conditions, the formulation of the question must be: Under which circumstances is participation worthwhile? And, if there are circumstances that favor participation, are these possibilities adequately exploited?

What might be the circumstances under which participation will show clear advantages? Three elements are relevant here. First, the situation should be of adequate complexity. If all the relevant factors for the decision are easily available, a single person can decide just as well as a group of persons – and with less effort for coordination. Second, the participating employees should have knowledge necessary to make decisions which is not available to others and which cannot easily be otherwise found. Third, the intended decision directly affects the working conditions of the employees. If this is not the case, the request for cooperation could easily be regarded as a burden or purely "pseudo-participation" with the consequence that the willingness to cooperate may diminish. The next section examines internal projects, measured by these elements, to explore the form and effect of the participation of employees.

6.3 Employee Participation and Success in Techno-Organizational Innovation Projects

In the ISI manufacturing innovation survey, a comprehensive set of questions covered the topic of internal project management. The participating manufacturers were asked to select a recent out techno-organizational project (e.g. the implementation of computer-supported material resource and production planning and control, the realization of computer integrated manufacturing or the implementation of working in teams). The project was to be named and then described according to a number of items. Some 674 manufacturers – that is 52 percent of the respondents – gave the respective information. This covered a large spectrum of techno-organizational process innovations, although it does not necessarily provide a representative selection of such projects in industry.

In 551 cases – that is more than four fifth of all recorded projects – project teams were formed so that representatives of different departments were directly involved in the planning, decision-making and implementation. Figure 6-2 shows that the main body of the team consists of representatives of the general management, the management of departments concerned and the staff of central organizational respectively electronic data processing (EDP) departments. These staff groups are to be found in 80 percent of the project teams and virtually always take over the project manager function. Only in every second project do affected employees belong

Figure 6.2   Participation of different staff groups with techno-organizational innovations

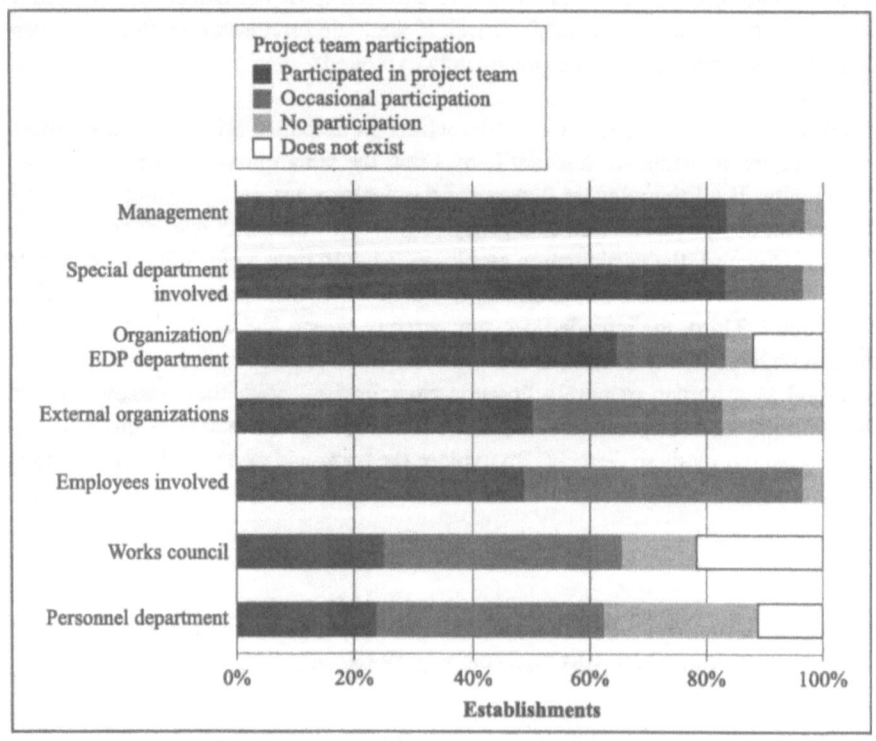

to the project team. In all other cases they are at least informed or occasionally consulted.

The workers' council is involved in these internal projects to a much lesser degree, even though the Betriebsverfassungsgesetz provides a number of codetermination rules or at least consultation rights related to internal decision making. Nevertheless, every fifth project team had representatives from the workers' council (a figure that will be striking to foreign observers from countries without codetermination). However, it should be mentioned that the function as a member of the workers' council did not automatically mean inclusion in the project team. Participating members of the workers' council usually had a duplicate role – as affected employees or employees with special competence. In one fifth of the projects, participation in or consultation of the workers' council was not possible as a workers' council did not exist in the company.

In all of the projects, the management strategy of the responding manufacturers provided for the implementation of a project team in four fifth of the cases and in approximately half of the cases, the participation of employees concerned in this team. However, strong differences between various project types in the use of project teams and in employee participation are apparent, based on the objective of the project (figure 6-3). Investments in manufacturing equipment as well as in the installation of computer-aided design (CAD) are most frequently handled without the implementation of a project team. However, if a project team is formed, the employees concerned take part in it more frequently than on the average. In particular, the future users of CAD contribute to the planning and implementation of almost eighty percent of CAD projects where project teams are formed.

In projects that go beyond the individual work place and that have strong organizational implications or even directly aim at new organizational solutions, almost all manufacturers establish project teams. The participation of the employees in the teams seems mainly to be regarded as being worthwhile in reorganization projects (introduction of work groups, the segmentation of production).

For computer aided production planning and control, and the related recording of shop-floor-data, most manufacturers refrain from the involvement of employees or restrict themselves to information or occasional consultation. (Strategic projects involving ISO 9000 certification and the relocation of production comprised too few cases to be included in the diagrams illustrated in figure 6.3.)

Figure 6.3   Project teams and employee participation, by project types

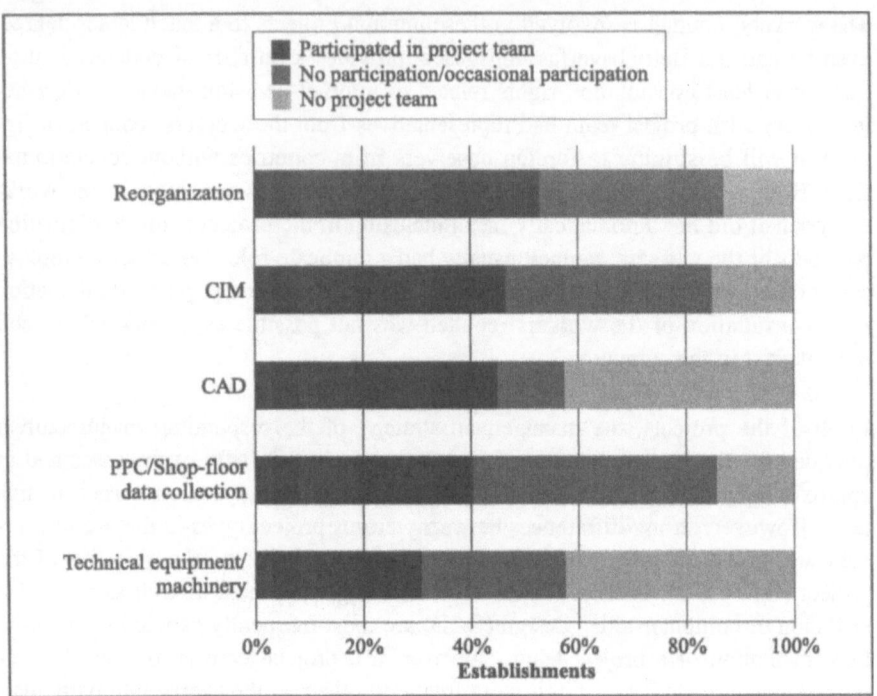

The way participation is handled also differs according to the various characteristics of manufacturers. Although smaller establishments rarely implement a formal project team, they more frequently report that they involve employees in project teams. An explanation for this difference in comparison to larger establishments could be that the employees are more acquainted with the entire operation of smaller manufacturers due to more transparency and less division of labor as well as an increased necessity to use their knowledge because of missing specialized departments. Their ability to contribute to decision-making may thus be higher. There is a close relationship between the type of project and the participation of employees in project teams. At the same time the distinguishable project types differ with respect to their content and objectives. There are small projects with a narrow spectrum of effects (e.g. the replacement of an old CNC-machine-tool with a new one) as well as complex projects (e.g. the introduction of production segments or of computer aided production planning and control).

As the implementation process of projects strongly depend on the project type, our analysis of the procedures and effects of projects controlled by employee participation focuses only on one project type. For this project type, we have selected the implementation of computer-aided production planning and control (PPC). PPC includes the use of materials requirements planning systems (MRP or MRPII), along with associated methods (such as shop-floor data collection).

Respondents to the survey report a large number (almost 200) of these projects, providing a good basis for analysis. For each project, manufacturers report on their implementation procedures and on whether the project's target was accomplished. This allows us to determine whether a project takes a different course if affected employees participate or not. It also allows us to assess how employee participation affects project outcomes.

In implementing PPC projects, respondents to the ISI survey reported four frequent problems that occurred in the course of the project.

- The project could not be carried out within the given time limit (time limit exceeded by more than 20 percent).

- The implementation of the project proved to be more costly than planned (requiring more than a 20 percent additional expense).

- The project concept developed in the beginning proved impractical during actual implementation.

- Project implementation met with resistance from the employees concerned or the workers' council.

More than one half of the PPC projects exceeded their initial time and expense budgets. Nearly one fifth of these projects ran into problems of impractical initial concept or employee resistance. Other difficulties in financing, inadequate supply of technology or poor management support from the management received few mentions.

Figure 6.4 shows how PPC projects with employee participation differ from those without with respect to the occurrence of difficulties. Surprisingly, in PPC projects in which employees participated in project teams, resistance was far more notable. Considering other important difficulties, the execution of projects with participation of employees, however, was better or just as good. In other project types (not shown in this figure), resistance was actually less frequent when the employees were involved in the project team. Resistance, for instance, in CAD-projects was noted in nine percent of the cases with participation of employees compared to 40 percent in projects without participation. Taking the average of all projects covered in the survey into consideration, the "acceptance rate" was slightly higher when the employees participated in the project team. However, the thesis of a widespread and significant positive impact on the acceptance of techno-organizational changes through the participation of employees concerned could not be proven.

Figure 6-4   PPC projects: Difficulties in implementation and employee participation

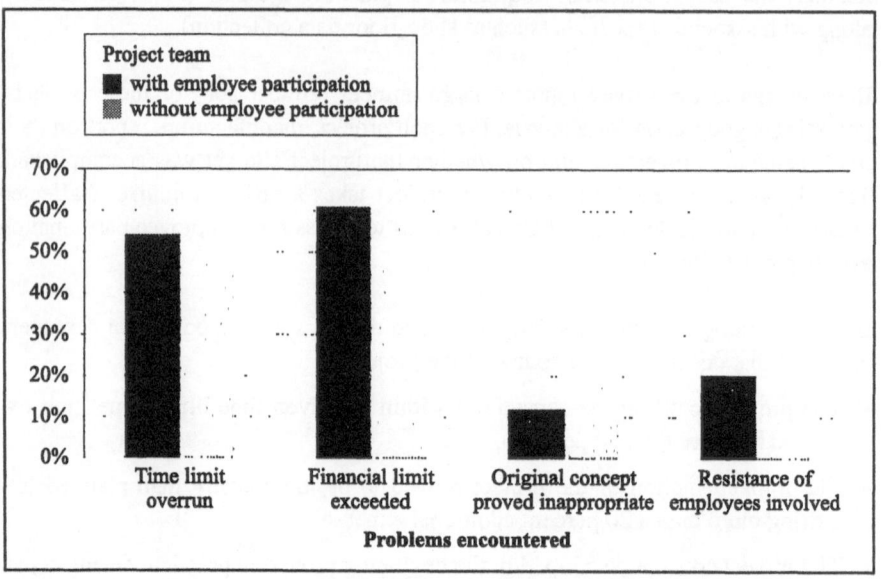

Why is the introduction of production planning and control systems more often confronted with resistance when employees are represented in the project team when, *prima-facie*, it might be anticipated that participation should actually help increase acceptance? There seems to be a complex cause and effect mechanism concerning the participation of employees concerned, leading to three complementary possibilities as explanations:

(1)   The expectation of reduced resistance through participation assumes that prior project planning offers a fully developed concept that is easily implemented. Participation then only has the purpose of helping in implementation without giving the employees concerned any real influence. Conflicts are foreseeable in this situation. If the employees have influence on the execution of the project through their participation and therefore take over some of the responsibility, they will only do so if their demands and proposals are given adequate consideration. Without a margin for this, an attempt to block the project implementation would be made. Thus, through participation, resistance can become even stronger if the offer to cooperate is not combined with the chance to share in decision making.

(2)   Furthermore, it can be assumed that participation of workers concerned also changes the quality of resistance. In the case of participatory decision making, a mere refusal to execute a project or a certain intention of doing so is not sufficient – it must be justified, that is to say, protest must be supported by ar-

guments. The project management has to reckon with sound criticism from the very beginning. In this way, both sides – management as well as employees – are forced to defend their point of view through argumentation. Through participatory decision-making, the exchange of views is encouraged, but this does not necessarily increase the acceptance of all decisions but rather only of those which are well founded. Others may even meet with less acceptance.

(3) The above considerations appear to apply particularly to the realization of computer-aided production planning and control. This is a complex process innovation which, at first glance, seems mainly technological, but which in reality is much more far reaching with multiple effects on work organization and working conditions. This is particularly evident if through a parallel implementation of electronic shop floor data recording, the activities of workers are more tightly monitored. In fact, PPC projects with data recording most frequently encountered problems and obstacles to implementation. These may not even have been noticed on the management level without the participation of the employees concerned in the project team.

The correlation between the occurrence of resistance and other obstacles (as shown in figure 6-4) confirm these considerations. Resistance of the affected employees then results in temporary delay of the project if there are no employees cooperating in the project team. If there is employee participation, resistance has much less effect on delays. Intensified discussions about the introduction and development of new solutions when there is resistance and at the same time the participation of employees in the project team has, however, frequently resulted in further tasks being defined. As figure 6-3 has already shown, participatory projects were more often planned in such a way that the original concept could be carried out. This indicates a better preparation of such projects. In fact, in the 21 projects in which the concepts developed in the beginning proved impracticable, resistance of the employees concerned was not mentioned as a problem in any of the cases. Probably, resistance in projects through participating employees in a positive kind of way does not principally aim at questioning the project as a whole but rather tries to make modifications of the project planning.

Difficulties in the course of a project are typical for complex projects. They are not necessarily an indication of bad work or lack of success. If the establishment of a project team aims at adapting planning to the specific frame conditions of the firm the difficulties occurring could be a sign that the necessary adaptation has been achieved. Such projects may in the end be more successful than those without adaptation. Besides the course to be taken by the project, the final assessment of success has therefore to be taken into account when evaluating the results of employee participation. Which differences can be determined in the project results of projects with and without participation?

It is not realistic to expect that a single written survey can capture the entire effects of a project on company performance. This is even truer given that very few manufacturers carry out an ex-post assessments of techno-organizational innovations. The manufacturers were therefore asked to assess how far the project achieved improvements for these four performance criteria: cost, quality, ability to deliver and flexibility. For these outcomes, projects with employee participation are slightly or even considerably better than those in which employees were not asked to participate in the project team (figure 6-5).

Anticipated cost and quality improvements in PPC projects were achieved about as frequently whether employees participated or not. For improvements in delivery and manufacturing flexibility, projects with employee participation performed noticeably better. Here, the higher goal attainment by manufacturers with employee participation is independent of establishment size or project characteristics such as volume or duration.

Figure 6-5   Goal attainment in PPC projects and employee participation

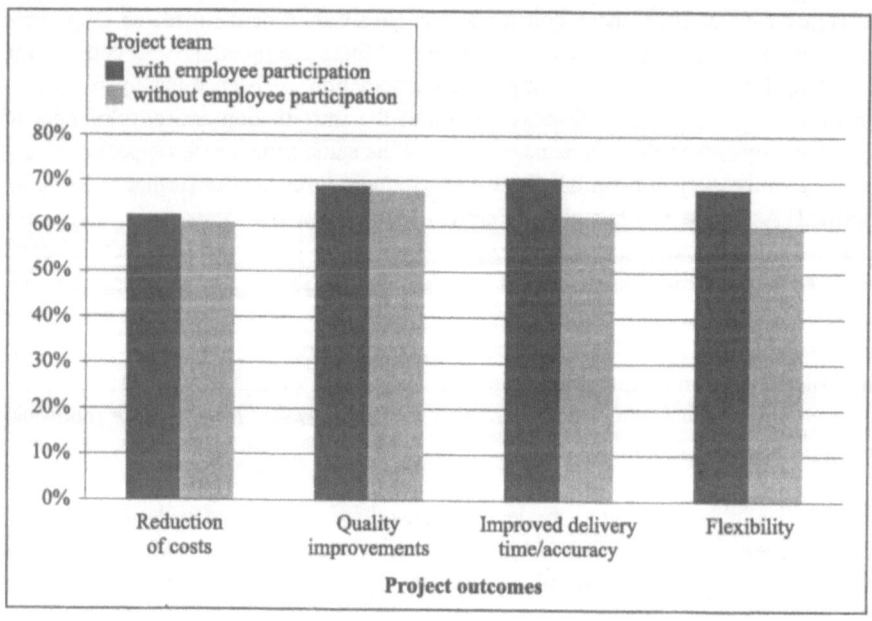

6.4   Conclusions

Germany has a long tradition of extensive participation of employee representatives in corporate decision-making. In recent times, new forms of direct participation and participation oriented organizational solutions have complemented institutionalized codetermination. On the European level, German companies are among the fore-runners in accepting employee participation, although they do not hold the leading position – for example, compare the study of the European Foundation (1997) or the figures for the diffusion in Sweden by NUTEK (1996) with the German situation presented here.

Any suggestion that increased participation of employees leads to disadvantages for German industry has not been proven by the results presented above from the manufacturing innovation survey of Fraunhofer ISI. Rather, companies with participation oriented organizational cultures show advantages in regard to relevant performance indicators. The advantages of participation of employees concerned also occur in projects that are aimed at realizing techno-organizational innovations, although the thesis that participation increases acceptance of the changes could not generally be proven.

However, the more that employees are valued as sources of important information and knowledge and the more their creativity is called for, then concerns about employee resistance become less relevant to the implementation of process innovations. In this case, employees are not just given the role to simply accept all procedures without discussion. In order to gain cooperative and responsible employees, one has to accept the possibility of resistance. This requires flexibility and openness to overcome conflicts that arise from resistance. The fact that projects with the participation of the employees more often accomplish their goals can be taken as an indication that the potential of participation of employees concerned has not yet been fully exploited. Particularly when the flexibility of production is the aim, increased involvement of the employees during the realization of the project counts.

Although the competitive position of firms in the German investment goods sector is arguably weaker now than in the past, the problems of these firms cannot be attributed to employee participation. Indeed, it might even be argued from a business point of view that there is some more room for attaining advantages from an increased use of participation oriented organizational solutions and improved participation of employees in decision-making.

6.5   Bibliography

Casson, M. (1994), Why are Firms Hierarchical? *Journal of the Economies of Business*, 1, pp. 47-76.

European Foundation for the Improvement of Living and Working Conditions (1997), *New Forms of Work Organisation, Can Europe Realise its Potential? Results of a Survey of Direct Employee Participation in Europe*, Office for Official Publications of the European Communities, Luxembourg.

Kirsch, W., Esser, W., Gabele, E. (1979), *Das Management des geplanten Wandels von Organisationen*, Stuttgart.

NUTEK (1996), *Towards Flexible Organisations*, Swedish Board for Industrial and Technical Development, Gotab, Stockholm.

Ritter, A (1992), Qualitätszirkel als Instrumentarium partizipativer Arbeitsgestaltung. In: Bungard, W. (Ed.), *Qualitätszirkel in der Arbeitswelt. Ziele, Erfahrungen, Probleme*, Göttingen, Stuttgart, pp. 37-50.

Schanz, G. (1992), Partizipation. In: *Frese, E. (Ed.), Handwörterbuch der Organisation*, Stuttgart, pp. 1901-1913.

Staudt, E. (1986), Innovation durch Partizipation: Möglichkeiten und Grenzen von Qualitätszirkeln. In: Staudt, E. (Ed.), *Das Management von Innovationen*, Frankfurt-Main, pp. 469-481.

7 Flexibility at Work

Gunter Lay and Claudia Mies

7.1 Introduction

Germany has an extensive array of rules and norms that regulate employee working hours and business operating times, including when during the week and weekends that manufacturers are permitted to operate. These limitations on working times are often highlighted in discussions about the barriers that constrain German industrial competitiveness. Indeed, the organization of working hours in industry can be viewed as a major indicator of rigidity or flexibility in the German manufacturing system and of the adaptation by German enterprises and workers to broader economic and societal changes. Fundamental changes in working hours in Germany involve negotiation processes on at least four levels: legal regulations, industry-wide collective bargaining agreements between trade unions and employers' associations, plant agreements and individual job contracts (Bullinger 1995).

Previously, efforts to modify working hours in Germany centered on the length of working hours (a 35 hour week, phasing out overtime in order to reduce unemployment) and on the implications of more varied working hours (health hazards from shift-work and night work versus better exploitation of manufacturers' capital equipment). Now, debate is concentrated – under the rubric of *flextime* – on another aspect: the distribution of working time as a function of the amount of work needing to be done (Hegner et al. 1995, Lehndorf and Bosch 1993, Seifert 1995).

There are two central issues in the current debate about flextime. The first concerns the period of time over which the flexible working hours have to balance out. What room for maneuver should there be when allocating different quantities of working hours to days of the week, weeks of the month or months of the year, in relationship to the amount of work that needs to be done? The second issue is about the assignment of authority to set actual working times within a frame of a flexible work times concept. How can a balance be struck between manufacturers' efforts towards flexibility and the preferences of individuals?

In the ISI manufacturing innovation survey, we were able to gather information on practices and trends in working hours in the German investment goods industry. This chapter describes some of the most significant results obtained on this aspect. There is a review of the diffusion of flexible working time concepts and an analysis of the use of various flextime models by plant size, industrial branch, business orientation, and type of production. The chapter then considers the link between flextime and the adoption of other changes in work organization within the plant.

7.2 Diffusion of Flexible Work Time Concepts

Flexible work time concepts in which daily working hours are not rigid are now being practiced by 56 percent of establishments in the Germany investment goods industry. Several different models are used, with variations according to whether the working time account has to balance monthly, yearly or over the working life of the employee. About 35 percent of manufacturers are using a model in which plus or minus hours have to balance out by the end of the month, while 33 percent of manufacturers are using a concept in which the work time has to balance out within the year. Only one percent of manufacturers use a model based on the employee's working life as the balance period. Therefore, in spite of prominent examples like Hewlett-Packard (see Pfander 1995) this model is not widespread in German industry.

It is clear from the percentages given above that some manufacturers have introduced more than one flextime model in parallel. Moreover, even among those firms that have adopted flextime, the intensity and universality of use varies. Of the manufacturers in the survey that have introduced a flextime model based on a monthly balance period, 16 percent are still in a pilot phase, 54 percent offer this model in some parts of the plant, and 30 percent are using it for all their employees. For manufacturers that have implemented flextime on a yearly balance period, 18 percent are still experimenting with the model on a pilot basis, 36 percent are offering it in some parts of the plant and 46 percent state that they are using it for all their workers. It is interesting to note that flextime models based on a yearly period tend to be used throughout the manufacturer more often than models based on a monthly period. (see figure 7-1, which illustrates the diffusion of flextime as well as the use of other flexible work concepts.)

Figure 7-1   Diffusion of flexible working time concepts in the German
                    investment goods industry

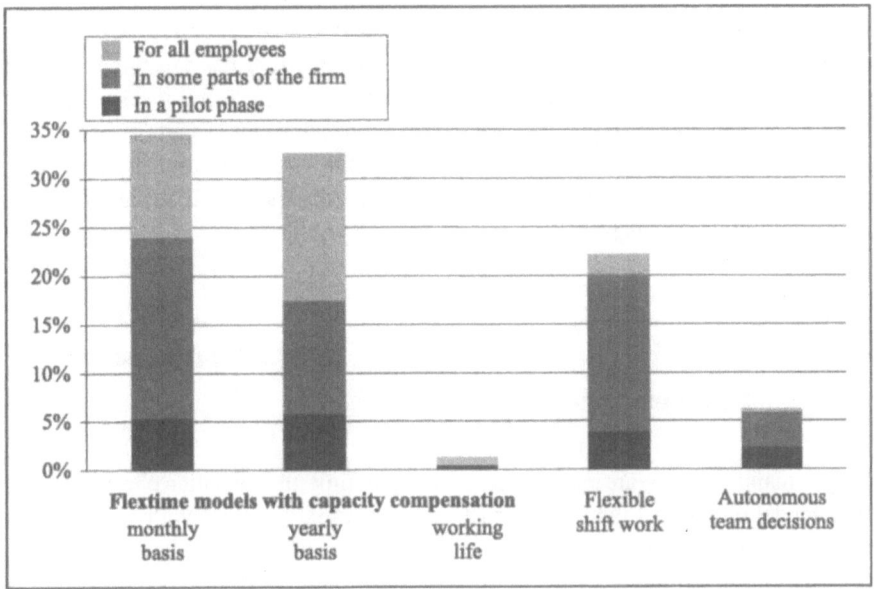

In the case of manufacturers that have implemented a working time model based on the employee's working life, 61 percent are using the model throughout the manufacturer, 31 percent are engaged in pilot schemes and 8 percent have introduced the model in parts of the manufacturer. However, due to the extremely small numbers of cases involved, these percentages should be interpreted very cautiously.

Experience with flextime has shown that manufacturers are confronted with problems particularly when this is combined with shift-work. With shift-work, the possibility of temporarily extending daily working hours to the legally permitted maximum of 10 is constrained by the fact that the same equipment may be needed by workers on another shift. This makes it all the more remarkable that 22 percent of manufacturers in the sample stated that they have a flexible shift-work model. However, in the great majority of these (73 percent) only some of the shift-workers were included in the model. Just 9 percent of manufacturers surveyed stated that they had flexible shift-work throughout. The remaining 18 percent are still gathering experience with pilot schemes for the realization of flexible shift-work models (figure 7-1).

In flextime models based on a monthly or yearly balance period, the decision on how daily work is to be organized can be made in several different ways. One option is to delegate the planning of working hours to work teams, who make autonomous decisions depending on the requirements of customer demands. This flextime concept, in which the regulation of work times is strongly decentralized and thus requires a high degree of competence and responsibility from employees, is implemented in 6 percent of establishments. One third of these manufacturers (37 percent) are currently trying out the feasibility of this concept in pilot schemes; however, more than half state that they have already introduced it in parts of production. At 6 percent, comprehensive implementation of this model throughout the plant is encountered just as seldom as manufacturers that are organized entirely on teamwork principles (see chapter 2).

7.3   Use of Flextime by Establishment Size, Industry and Strategic Orientation

Larger manufacturers are more likely to use flextime than smaller ones. The work time model based on a monthly balance period is used in 30 percent of small manufacturers (up to 100 employees), in 41 percent of medium sized manufacturers (100 to 500 employees) and 45 percent of large manufacturers (more than 500 employees). The flextime model that takes the year as a balance period is being used in 22 percent of small manufacturers, 47 percent of medium-sized manufacturers and 54 percent of larger manufacturers (figure 7-2). Large and medium-sized manufacturers thus prefer the "annual" model. For flexible shift-work, use also varies by establishment size. Flexible shift-work has been implemented by 17 percent of small manufacturers, 25 percent of medium sized manufacturers and 46 percent of large manufacturers.

These values may reflect the difficulty small and medium sized manufacturers have in finding the resources for planning and introducing flextime concepts, as well as the work involved in administering systems of this kind. It should also be borne in mind that know-how about flextime options is more likely to be present in larger manufacturers.

There are also variations in the use of flextime by specific industries within the investment goods sector, although no clear trends emerge. Compared against the averages for the whole investment goods sector, electrical engineering manufacturers are more likely to use monthly flextime concepts (39 percent) and autonomous team decisions (11 percent). On the other hand, in the automotive industry, use of the annual model is above average (38 percent). In the mechanical engineering industry, the distribution of the various flextime concepts closely matches the average for the investment goods sector as a whole. In the other branches of the investment

Figure 7-2   Use of flextime models in the German investment goods industry, by establishment size

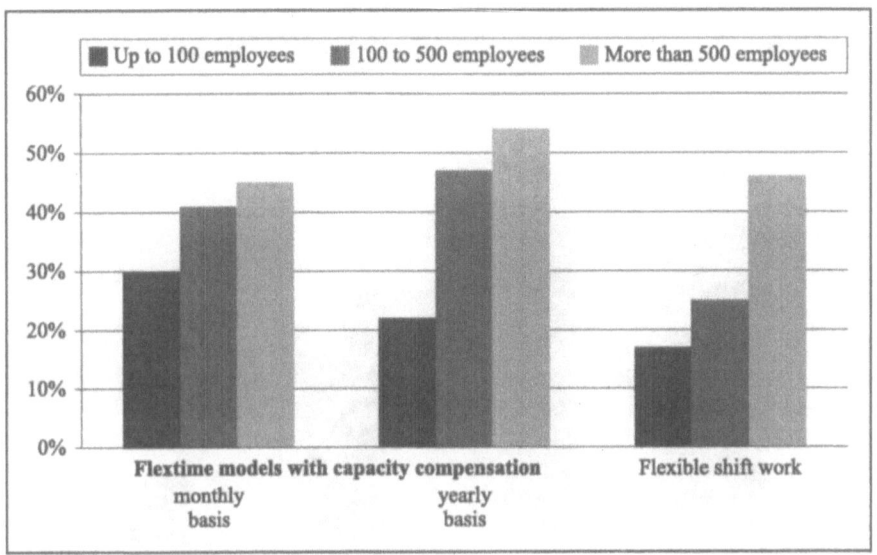

goods industry, the flexible organization of shift-work (at 26 percent) is noticeably more diffused than in the other investment goods subsectors. We will return later to the question of how far these differences are attributable to specific production conditions and needs.

We also probed the extent to which the use of flextime models is related to manufacturers' strategic business orientation. In the ISI survey, we categorized manufacturers by four strategic orientations (on the basis of their self-assessments):

- manufacturers that want to compete on the market mainly via their product prices, and are thus primarily interested in reducing their costs;

- manufacturers that are pursuing a quality-oriented strategy by offering very high quality products in "up-market" price segments;

- manufacturers that are trying to distinguish themselves from their competitors by their broad range of products; and

- manufacturers that aim to achieve success through fast and flexible delivery.

Figure 7-3   Use of flexible work times in the German investment goods in-
                  dustry, by competitive strategies

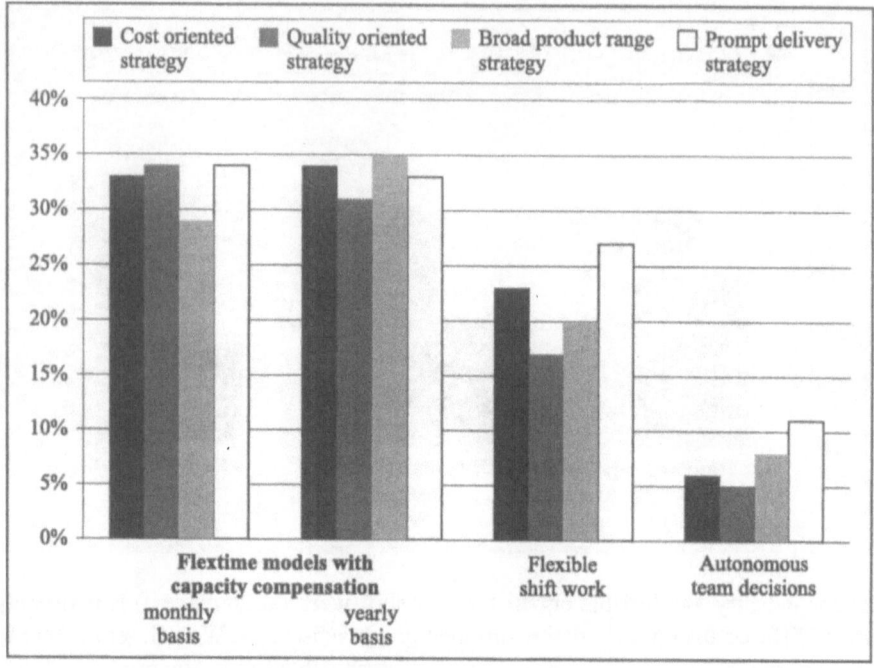

We then compared the use of flextime models with these different strategic orientations (see figure 7-3). There is a relatively high use of flexible shiftwork (27 percent
of the group) in manufacturers where improving the ability to deliver is the strategic
competitive aim. These manufacturers also are more likely to devolve flextime decisions to autonomous teams (11 percent of the group). This finding reflects the
special need for flextime by manufacturers who compete by prompt delivery. These
plants depend more than other establishments on the ability and willingness of
workers to work together to responsively meet customer demands.

7.4   Influence of Product and Type of Manufacturing on the Implementation of Flextime Models

It is possible that the differences in the diffusion of flextime models with the investment goods sector reflect different operational requirements for working time
concepts resulting from the type of products manufactured and the method of pro-

duction. This section examines the diffusion of flextime models as a function of four factors:

- the complexity of the products manufactured;
- the type of production – production program (producing standard customer orders or to stock) or custom-made products;
- batch sizes;
- the organizational principle of production (job shop production, flow shop production, or manufacturing segmentation).

Based on this analysis, three types of manufacturers can be distinguished (figure 7-4). Each type differs not only in the use of flextime but also in their manufacturing and product characteristics. It appears that the manufacturers that have shown least interest in flexible working time models are those with predictable sales expectations able to produce for inventory independently of individual customer orders. In the survey, this is clearly demonstrated by the below-average percentages obtained for flextime models in these manufacturers, not only for monthly flextime models (11 percent) but also for flexible shift-work (8 percent). In this group of manufacturers the necessity for capacity-related flexibility is - or is considered to be - so small that there is not so much pressure to introduce flexible working hours.

Figure 7-4   Use of flextime by differences in product and manufacturing characteristics

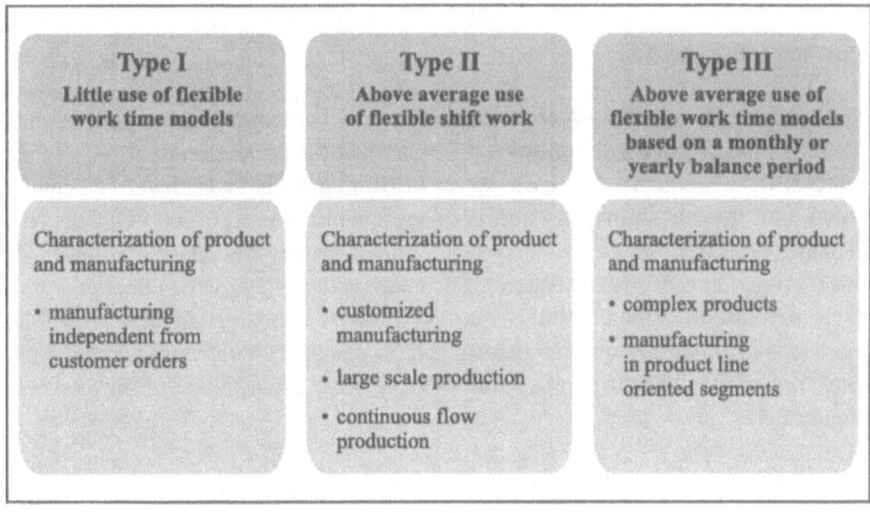

If manufacturers follow a fixed production program but also carry out customer-specific variants within this program, and if the products concerned are also simple, produced in large batches and in flow shop production, they have – understandably – a higher than average tendency to adopt flexible shift-work. In this situation, meeting customer-specific wishes for variants causes fluctuations in capacity that have to be accommodated despite a large-scale flow shop production.

Where manufacturers of complex products are not organized in job shop or flow shop production, but have already regrouped their machines in product or customer-oriented work cells (manufacturing segmentation), flextime models tend to be implemented in order to accommodate fluctuations in demand. This is clearly shown by the fact that product-oriented segments are linked with yearly- or monthly-based flextime models in 40 and 38 percent of cases respectively.

7.5   Influence of Work Organization on the Use of Flextime Models

New production concepts involve changes in work organization that include the decentralization of authority to decide, task integration and teamwork. Flexible models of working time have often been viewed as measures to support these other fundamental changes in work organization (Mies 1997 p. 184). But what exactly is the correlation between the implementation of new forms of work organization and the use of flextime models? The results from the ISI survey can be used to provide answers to this question.

The analysis shows that where changes in work organization are implemented, flextime models are used more often. In manufacturers with decentralized decision processes, the frequency of yearly-based flextime models (44 percent of manufacturers) and flexible shift-work models (32 percent of manufacturers) is higher than average. Working time decisions by autonomous groups are also more widely diffused (10 percent of manufacturers). For manufacturers that have integrated previously separate functions, monthly- and yearly-based flextime models are slightly more frequent than average (39 percent and 35 percent respectively) and the implementation of flexible shift-work models (31 percent) is considerably above average (figure 7-5).

In manufacturers using teamwork, the diffusion levels for flextime models based on a monthly or a yearly period (43 and 38 percent of manufacturers, respectively), for flexible shift-work (27 percent of manufacturers), and for autonomous group decisions (10 percent of manufacturers) are all higher than the corresponding figures for manufacturers where teamwork is not used. (See figures 7-6 and 7-7.)

Figure 7-5   Flextime and decentralization of planning and control functions
in the German investment goods industry

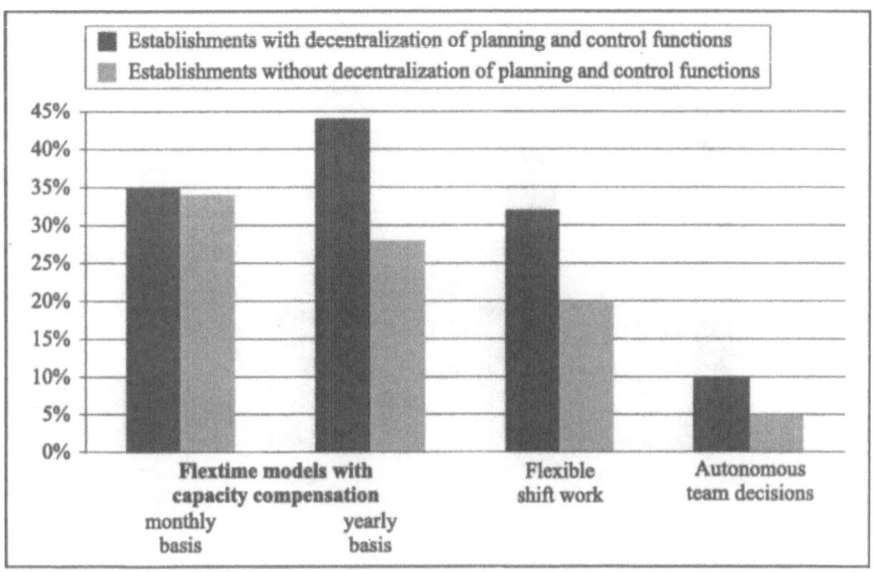

Figure 7-6   Flextime and task integration in the German investment goods
industry

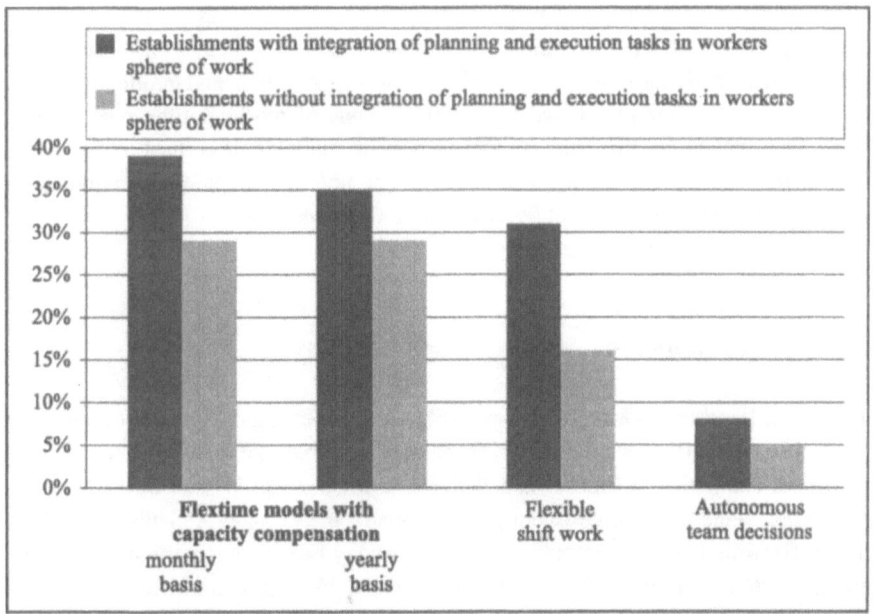

Figure 7-7   Flextime and the use of teamwork in the German investment
              goods industry

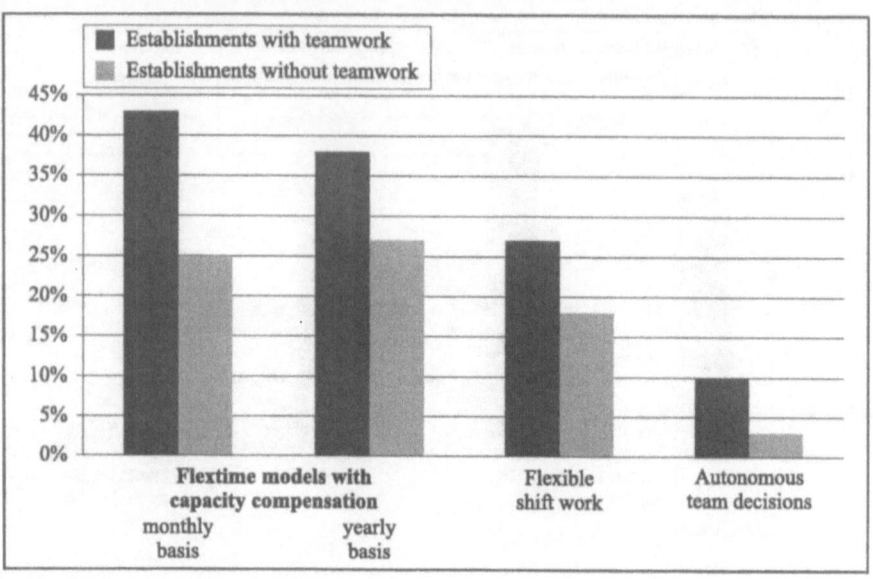

These correlations indicate that between adopters and non-adopters of the organizational elements of new production concepts in manufacturing, there are differences
in use of flextime concepts. But these differences are not so great as would have
been expected from the discussion on the mutual interdependency of these two organizational fields (see Schneider 1992). New organization and production structures are not yet complemented by flexible forms of working time over a broad
front. One reason for this may be that until now manufacturers have concentrated
their efforts on re-organizing work structures, and that a parallel adaptation of
working time models could not be managed at the same time.

7.6   Conclusions

The use of flexible working times - with the aim of meeting customers' orders faster
while also reducing costs due to overtime pay, and then compensating by reducing
personnel capacities at times when demand is lower - has passed through the experimental stage in the investment goods industry in Germany. For many manufacturers, flextime has now become a part of everyday life. This is particularly evident
among large manufacturers. But it is even more evident in manufacturers with custom-made production involving many variants than for manufacturers with large

batch production programs. Additionally, flextime is found more frequently in manufacturers that have introduced new organizational production concepts, such as product-oriented manufacturing segments or teamwork, than in manufacturers that (still) adhere to traditional modes of production.

It is interesting to note that flexible working hours have been accepted more readily and with much less controversy by German industry than by the country's service and retail sectors. Industry is clearly more exposed to international competitive pressures that demand greater flexibility, whereas the service and retail trades are more domestically oriented. Yet, there are other structural factors involved. In industry, capacity demand differences do not occur at such short notice, periods of low and high capacity demand are longer and the compensation of peak capacities can be more freely planned within the working day or the working week. This planning freedom can be used to coordinate the interests of the manufacturer and the individual in such a way that both sides can benefit from flexibility.

The delegation of decisions on work times to autonomous groups represent a potentially effective way of allowing flexible, decentralized organization of work by the employees themselves. However, the fact that this combination of flextime and devolved workgroup decision making is still not very widespread as yet should motivate manufacturers to think even harder about how the flexibilization of working hours can be linked to other improvements in the workplace.

7.7   Bibliography

Bullinger, H. (1995), *Arbeitsgestaltung. Personalorientierte Gestaltung marktgerechter Arbeitssysteme*, Teubner Verlag, Stuttgart.

Hegner, F., Bittelmeyer, F., van Bruggen, W., Heim, G., and Kramer, U. (1995), *Erfolgsfaktor Zeit*, Arbeitgeberverband Gesamtmetall, Köln.

Lay, G. and Mies, C. (Eds.) (1997), *Erfolgreich reorganisieren - Unternehmenskonzepte aus der Praxis*, Springer Verlag, Berlin and Heidelberg.

Lehndorf, S. and Bosch, G. (1993), *Autos bauen zu jeder Zeit - Arbeits- und Betriebszeiten in der europäischen und japanischen Automobilindustrie*, Edition Sigma, Berlin.

Mies, C. (1997), Der Kundenauftrag bestimmt den Feierabend. In: Lay, G. and Mies, C. (Eds.), *Erfolgreich reorganisieren - Unternehmenskonzepte aus der Praxis,* Springer Verlag, Berlin and Heidelberg, pp. 183-205.

Pfander, W. (1995), Das Hewlett-Packard Arbeitszeitansparmodell. In: Wagner, D. (Ed.), *Arbeitszeitmodelle*, Verlag für angewandte Psychologie, Göttingen, pp. 173-182.

Schneider, D. (1992), Lean Production – Herausforderungen für die Gestaltung der Arbeitszeit, *Personalführung*, 9, pp. 698-707.

Seifert, H. (1995). Spielräume für Betriebsnutzungszeiten, *WSI*, 10, pp. 641-645.

Wagner, D. (Ed). (1995), *Arbeitszeitmodelle*, Verlag für angewandte Psychologie, Göttingen.

8 Innovation versus Emigration - New Production Concepts and Transborder Relocation

Steffen Kinkel

8.1 Introduction

An intense debate is now underway about the advantages – and especially the disadvantages – of Germany as a location for industry. In particular, criticism is leveled at Germany's unfavorable framework conditions such as high taxation, excessive bureaucracy, numerous regulations, and high wages and wage-related costs. At the same time, German enterprises increasingly seek to expand their sales by opening up new international markets and fields of business close to foreign customers. Thus, whether to lower costs or to serve international markets, many German firms have relocated – or are considering the relocation of – parts of their production abroad.

This chapter uses data from the 1995 ISI manufacturing innovation survey to investigate current trends in relocation and to investigate the effects associated with transferring production to locations abroad (see Kinkel 1996). Two major themes are examined. First, the chapter assesses the transfer of production capacities to locations outside Germany in the previous two years and examines plans to relocate production abroad in future years. Second, the chapter explores the relationships between the relocation of production abroad and the business performance of production functions remaining in Germany, focusing on flexibility, productivity and the time to market of new product development.

8.2 Extent of Actual and Planned Relocations

Of the investment goods establishments questioned in the ISI survey, 17 percent stated that they had relocated production capacities abroad in the last two years. Of these, more than four-fifths are planning more production transfers in the coming two years, while 20 percent of the remaining establishments want to take this step for the first time. This means that 31 percent of establishments are planning first or additional transfers abroad in the next two years. If these establishments are differentiated according to structural characteristics, the following picture emerges (figure 8-1).

Figure 8-1   Actual and planned production relocations abroad, investment goods sector

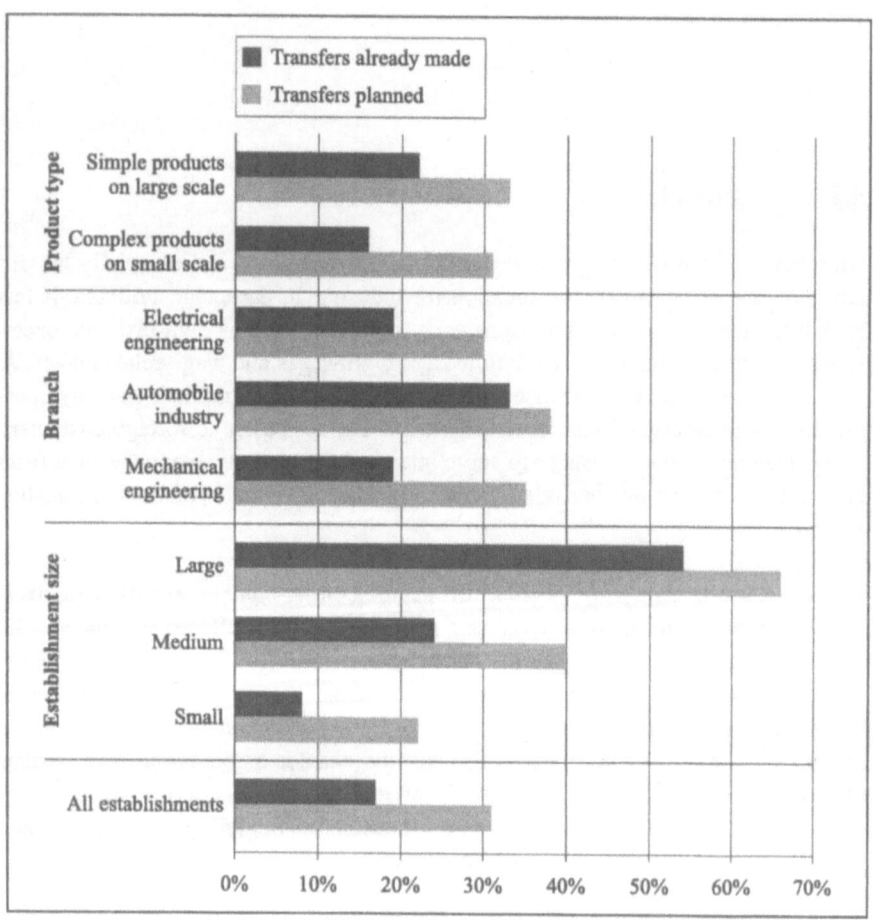

In the past, production transfers have been particularly prevalent for large manufacturers with 500 or more employees. More than half the establishments in this category (54 percent) have relocated production capacities abroad over the last two years. On the other hand, small manufacturers with less than 100 employees (8 percent) and medium-sized manufacturers with a workforce of 100 to 500 (24 percent)

have not been nearly so involved in relocations so far. However, a "catching-up" effect can be observed. If their plans are realized, the relocation rates for small and medium sized manufacturers will double or triple in the next two years, compared to the previous two-year period.

The automobile industry, where one-third of the establishments in the survey have already made production transfers abroad, is further along in this respect than other sectors such as mechanical engineering and electrical engineering (each with transfers made in 19 percent of establishments). For the coming two years, however, these sectors show nearly the same level of planned transfers as the automobile industry. The mechanical engineering industry reports a relative relocation growth rate – compared to the past two years – of 84 percent (a jump from 19 percent to 35 percent of establishments contemplating relocations abroad). In electrical engineering, the comparable relocation growth rate is 58 percent (from 19 percent to 30 percent of establishments).

The types of products most often mention as particularly suited for relocating production abroad are those which combines a low degree of technological complexity with a high degree of standardization, and can thus be produced on a large scale. Routine products of this kind can be manufactured in labor-intensive or automated processes, depending on the type of product and the production location. Due to their relative simplicity, this group of products can also be produced by less skilled or qualified personnel than are required for other product types. Indeed, analyses show that producers of simple products in large-scale production are more likely to have made production relocations (22 percent of establishments, compared with 16 percent in the case of producers manufacturing complex products in smaller-scale production). However, a "catching-up" effect seems to be setting in. For transfers planned for the next two years, a significant difference between these two categories is no longer apparent. This is an important shift, suggesting that more foreign locations have developed the capabilities to sustain complex production.

8.3   Production Relocation and Strategic Orientation

When making decisions about transferring production abroad, a firm's strategic orientation is decisively important. On the one hand, relocation may be used, in a cost-oriented strategy, to realize lower production costs, thus gaining competitive advantages in the price competition. On the other hand, establishments aiming at shorter lead and delivery times, or the rapid commercialization of innovative products, in a performance-oriented strategy, may be enabled to exploit new potentials in these performance parameters by relocations that bring them "closer to the market."

Evidence from the ISI survey allows us to assess the relative importance of these two strategies (see also figure 8-2). Manufacturers whose primary aim is to become or remain competitive in price competition by pursuing a cost-oriented strategy transfer production capacities abroad more often (19 percent) than establishments whose strategy is to distinguish themselves through performance parameters (13 percent). Cost-oriented are presumably attracted by the lower costs of production outside of Germany. However, establishments with a performance-oriented strategy, arguably do not as strongly perceive that the advantages of being closer to foreign markets outweigh the disadvantages of supplying those markets from Germany.

The strategic orientation of establishments has a stronger influence on relocation decisions than do basic frame conditions within the firm, such as the complexity of products or processes. The percentage of production relocations abroad in establishments with a performance-oriented strategy turns out to be as low as 13 to 14 percent, irrespective of whether the firm manufactures simple products on a large scale or complex products on a smaller scale. A different picture emerges for establishments with a cost-oriented strategy: here the 25 percent of establishments that have made production transfers are mainly large-scale producers of simple products. It seems that these firms react particularly strongly to pressures to reduce costs by producing abroad.

Figure 8-2   Production transfers already made, differentiated by strategic orientation and product complexity

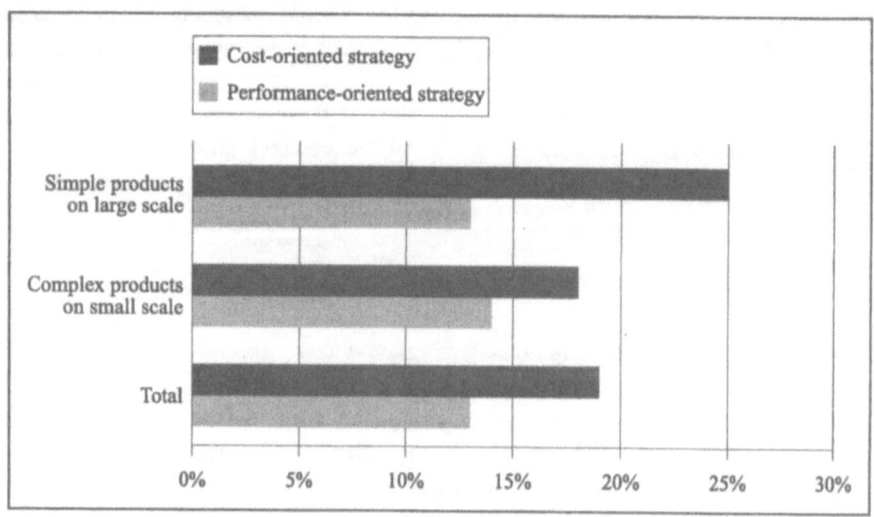

8.4   Production Relocation and Performance Parameters

To analyze the effects of production transfers on a firm's performance parameters, such as flexibility or time to market, it is necessary to clarify the various ways in which areas of production can be transferred to locations abroad. Full relocation may mean that whole product lines are transferred to foreign locations. Partial relocation may that preliminary, intermediate or finishing stages of a product line are relocated abroad. This extension of the production chain generates additional two-way export-import interfaces, and creates a need to co-ordinate storage capacities and the timing of materials and deliveries with product orders.

The interface problems associated with the partial relocation of product lines abroad make it relevant to compare overall lead times as a criterion for establishments' flexibility in reacting to customer orders. This comparison reveals a significant difference: establishments that have made production relocations abroad within the last two years have average overall lead times of 80 days, as opposed to 61 days for establishments whose production is located exclusively in Germany. In other words, establishments that have relocated production have their lead times that are 31 percent longer (figure 8-3).

Producers who manufacture complex products on a relatively small scale and who relocate production abroad do so with a sacrifice of flexibility. At 125 days, the overall lead times of these establishments are an average of 39 percent longer than in establishments of the same type which have not transferred any parts of production abroad (average lead times: 90 days). This difference is not attributable to distortions caused by the parameter of firm size. The group of producers of complex products in small-scale production does not include a higher-than-average proportion of large enterprises, with their relatively long lead times and more frequent relocation of production capacities.

By contrast, producers manufacturing simple products on a large scale show no significant differences concerning flexibility. Their average overall lead-time lies between 22 and 23 days, regardless of whether they have made production relocations abroad or kept their entire production in Germany. For one thing, this group of establishments tends to transfer whole product lines, so that the problem of extended production chains requiring additional management only occurs to a limited extent. For another, it appears that the simpler product structure and the resultant simpler process structure in these establishments means that new production chain links are more easily mastered.

If a firm is pursuing a performance-oriented strategy with the aim of distinguishing itself from its competitors by becoming an innovation leader, the product development time (time to market) becomes one of the decisive criteria by which this strat-

Figure 8-3   Lead times and product development times related to production
relocation and product complexity

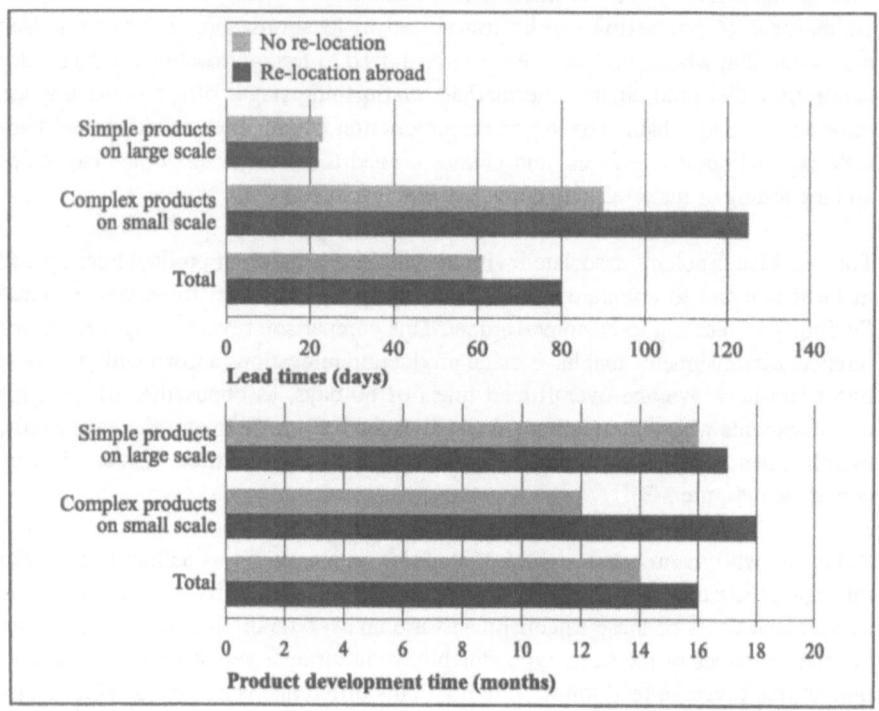

egy stands or falls. The shorter the time to market – defined as the elapsed time
taken to develop an innovative product to the point where it can be launched on the
market – the greater are the chances of success in competition. It has been shown
that relocating production abroad increases the average time to market by over 14
percent, from 14 to 16 months. However, there are also differences in the pace of
innovation depending on the complexity of the products being manufactured.

The negative impacts on lead-time of relocating production capacities abroad are
particularly apparent for complex products manufactured on a small scale. In this
situation, the average time to market goes up by 50 percent, from 12 to 18 months.
The additional interfaces created by relocation appear to inhibit product development processes, especially if the new developments concerned are complex products.

On the other hand, the average time to market for producers of simple products
manufactured on a large scale is between 16 and 17 months, regardless of whether
production relocations have been made or not. The simpler production structures

mean that development processes are easier to master and their efficiency hardly seems to suffer from the existence of extended production chains.

8.5   Production Relocations and Productivity

The main motive of producers of investment goods in transferring production capacities to foreign locations is to exploit cost reduction potentials. What effect does this cost-driven relocation have on productivity? The ISI survey allows productivity to be calculated in terms of value added per employee created within the establishment. For all establishments included in the survey, the average value added per employee was 126,000 DM.

Within this average, the relocation of production is associated with differences in value added per employee. It appears that establishments which have relocated some parts of their production abroad in the last two years have a value added per employee in Germany that is 18 percent higher, at 145,000 DM, than enterprises which do not produce abroad (123,000 DM). Thus at first sight it would seem that relocations are associated with better productivity performance at home. However, it is clear that in a strategy based solely on relocation, the average productivity of the parts of production remaining in Germany goes up. This is because less productive product lines or production processes have been rationalized by transferring them abroad (and so are no longer counted in purely German-based calculations of value added per employee). This does not imply that the processes remaining at home have been improved. Indeed, a critical question arises as to whether productivity might be even higher in a strategy that aims at optimization rather than at relocation.

One approach to improving the productivity of establishments in Germany is the implementation of new production concepts. As shown in Chapter 3, judicious use of new production concepts can substantially improve the productivity of establishments (see also Lay, Dreher and Kinkel 1996, Lay and Mies 1997). From the two alternatives of relocating production or modernizing production structures by the introduction of new production concepts, the following matrix of strategies can be derived (figure 8-4):

Figure 8-4   Production strategies distributed between relocation and
the introduction of new production concepts

Production Strategy	No use of new production concepts		Use of intrinsically compatible elements of new production concepts	
	Type	Description	Type	Description
No relocation of production	O	Traditionalist	B	Modernizer
Relocation of production	A	Relocator	C	Partial Modernizer

This matrix defines four strategies. *Traditionalists* (Strategy O) have neither introduced individual elements of the new production concepts, nor relocated parts of their production abroad in the last two years. *Relocators* (Strategy A) have made production transfers to locations abroad, but have not realized any new production concepts in the parts of production remaining in Germany. *Modernizers* (Strategy B) have introduced new production concepts at their German production sites, without relocating parts of production abroad at the same time. Finally, *Partial Modernizers* (Strategy C) have relocated production capacities abroad in the last two years and at the same time have also implemented new production concepts in the parts of production that have not been relocated.

The establishments responding to the ISI survey were allocated to this matrix, and the productivity levels under each strategy for the parts of production remaining in Germany were calculated (see figure 8-5).

The establishments with the *lowest productivity* are those which pursue Strategy O. These have not introduced any elements of the new production concepts, nor have they made any relocations of production. The average value added per employee for establishments following this strategy is only DM 99,800 per employee.

Higher productivity is attained by establishments following Strategy A, with an average value added per employee of DM 140,000 per year. In this variant, the areas of production that are least productive in Germany are relocated abroad, but no new production concepts are realized. The average productivity of the parts of production remaining in Germany then goes up by 40 percent, solely because the previously most "unproductive" manufacturing processes are no longer there. However, the processes remaining in Germany do not become more efficient due to the relocations.

Figure 8-5   Average productivity correlated with production relocation and
the use of new production concepts

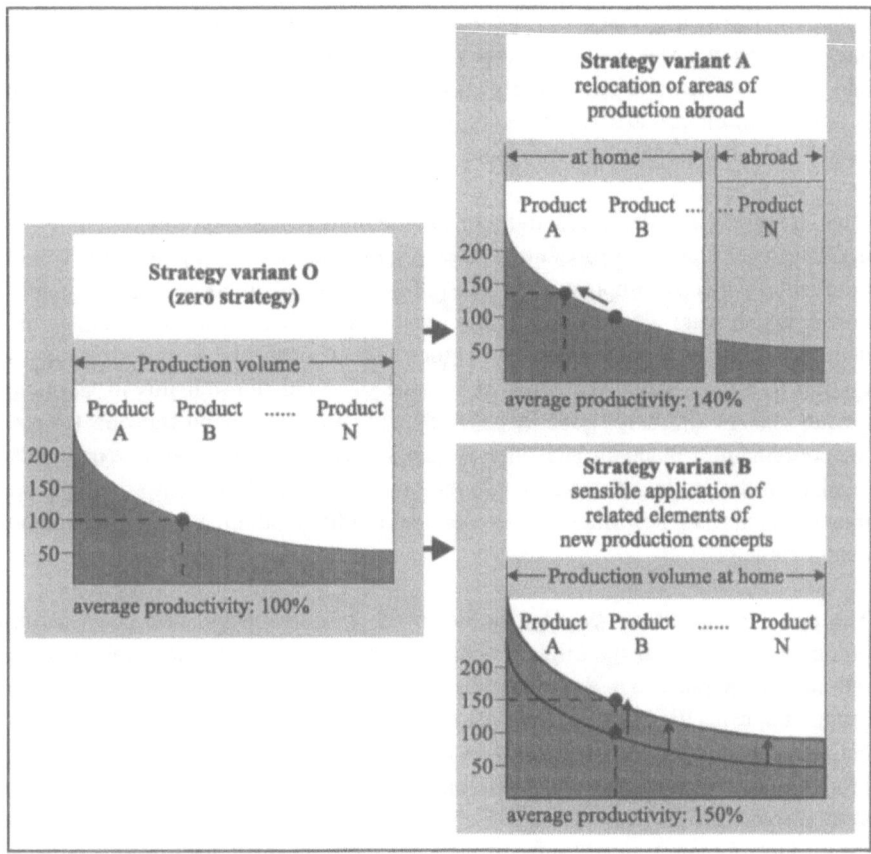

The establishments that show the *highest productivity* are the ones that have
adopted Strategy B, i.e. by solely implementing intrinsically compatible elements of
new production concepts at their location in Germany. These include, for instance,
segmentation of production supported by work reorganization with decentralized
decision-making competencies and broader job content, the combined use of various concepts for quality assurance, or intra-firm and inter-first just-in-time concepts. The use of intrinsically compatible elements of new production concepts improves the overall productivity of establishments by rendering much more effective
all product lines and areas of production in which the new concepts can be applied.
This boosts the average value added per employee by approximately 50 percent
over the base Strategy O, to an annual level of DM 150,000 per employee. However
– unlike strategy variant C – the entire production volume remains at home.

Surprisingly, in Strategy C, the combined use of strategies B and A, there is no summation of the productivity effects. In this variant, establishments transferred parts of production to foreign locations while trying at the same time to render the remaining German production more effective through the introduction of new production concepts. The average annual value added per employee is DM 145,000 or 145 percent of the base strategy. But the productivity of Strategy C establishments is lower than the productivity of establishments (Strategy B) which did not engage in any relocation but put all their efforts into the new concepts.

Thus, it appears that the productivity of the previously most unproductive processes in exclusively domestic production can be greatly improved by introducing the new production concepts. Improvement possibilities are high in comparison to a situation in which these areas of production are simply removed from the German productivity statistics ("swept under the carpet," so to speak), by transferring them to foreign locations. Moreover, when the whole of production remains in Germany, greater *synergy effects* are generated. When using new production concepts the need for coordinating, planning and other administrative tasks is to some extent independent of the production volume. On the other hand, when parts of production are transferred abroad, new interfaces are created requiring additional coordination and administration.

The productivity effects that can be achieved using compatible elements of new production concepts at the original "home" location can more than compensate the rationalization potentials that could be exploited by cost reduction in relocating production capacities abroad. Contrary findings only occur if firm types are differentiated according to the complexity of their products (see figure 8-6). The superior productivity potential of new production concepts compared to relocation, demonstrated above for the investment goods industry as a whole, also holds good for the sub-group of manufacturers of complex products in small-scale production. By contrast, where large-scale manufacturers of simple products have used intrinsically compatible elements of new production concepts as a rationalization instrument in exclusively domestic production (Strategy B), their productivity has been lower than in establishments that have also - or exclusively - realized cost advantages by relocating production capacities abroad (Strategies A and C).

The rationalization and cost reduction potentials of new production concepts, as measured by their contribution to improving productivity, still appear insufficient for establishments that produce simple, mass-produced products: the effects that can be achieved in solely domestic production are not substantial enough to constitute an alternative to relocation strategies. There would seem to be a need here for new concepts capable of being adapted intelligently by these establishments, thus enabling them, too, to stay competitive in the price competition while remaining in Germany.

Figure 8-6   Productivity related to production strategy and product complexity

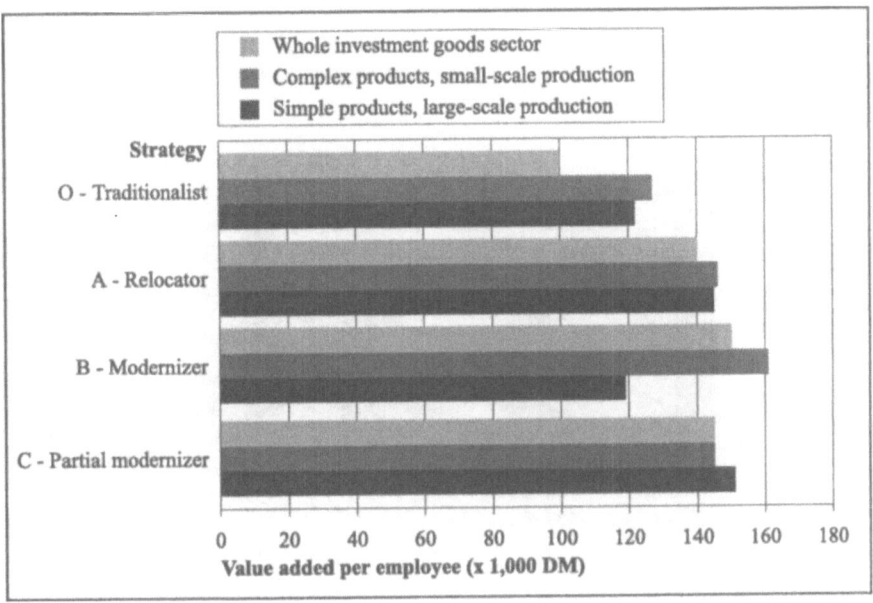

Another reason for the absence of rationalization effects is to be found in the strategic orientations of producers of relatively non-complex products in which the new production concepts are embedded. Considering the effects on productivity and flexibility of the use of complementary elements of new production concepts in different types of establishments, the following relationships are found (figure 8-7):

- In establishments that manufacture complex products on a relatively small scale, and have implemented complementary elements of the new production concepts, productivity is 34 percent higher than in establishments of this type that have not modernized their production appropriately.

- For producers manufacturing simple products on a large scale, "combined" use of these concepts brings a much smaller productivity advantage of six percent. On the other hand, with manufacturers of complex products, combined use of several elements has no significant influence on overall lead times.

- By contrast, the lead times of the firm type "large-scale producers of simple products" are reduced by 48 percent if different elements of new production concepts are judiciously combined. This group of establishments seems to use the potentials of the new concepts specifically to strengthen performance-oriented competition factors such as flexibility.

Figure 8-7   Productivity and lead times related to the use of new production
              concepts

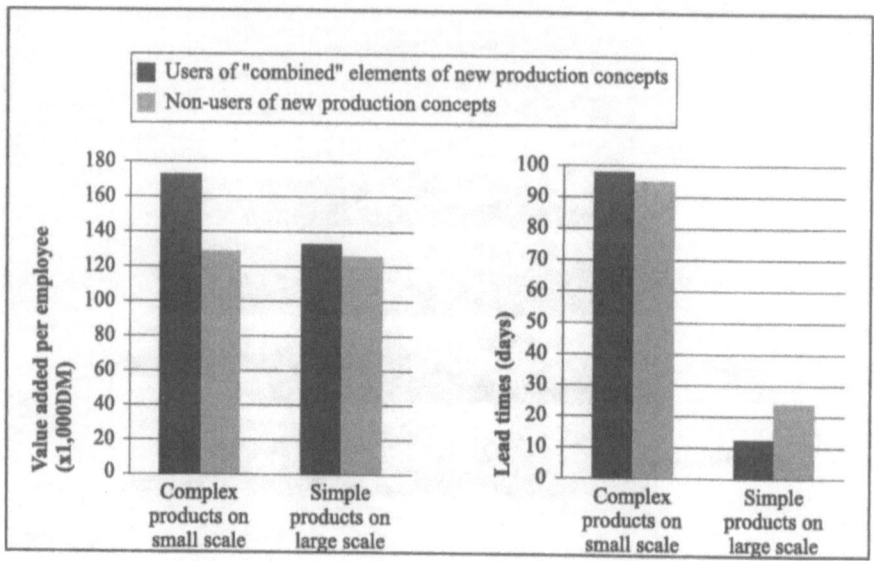

In this context, it becomes apparent that an "even" distribution of the effects achievable with new production concepts is only possible up to a point (see Lay 1997). If efforts are concentrated on improving performance-oriented parameters, rationalization is only possible to a limited extent. If, as seems to be the case with producers of complex products, all the firm's efforts are put into realizing the cost reduction potentials of reorganization, then no contributions can be expected to improving other relevant competition factors, apart from the product price.

8.6   Conclusions

In the German investment goods industry, the trend towards relocating production capacities abroad will accelerate in the immediate future. Almost one-third of enterprises are planning (first or further) relocation activities in the near future, as compared to the 17 percent of establishments that have made relocations in the recent past. Particularly in small and medium sized establishments, a very strong "catching-up" effect can be observed. If these plans are realized, relocations will more than double in the next two years. Increasingly, these enterprises appear to be experiencing the possibility of internationalizing their business activities as a pressure that also affects their production, compelling them to produce in other locations.

The strategic orientation of establishments has a more sustained influence in this respect than characteristics such as firm sector or product complexity. Production capacities are being relocated particularly by establishments that are seeking to strengthen their cost factor in competition via low-cost production abroad. Establishments with a strategy based on high performance, on the other hand, tend not to relocate so often, for fear that some relevant performance parameter might be weakened. In fact, results indicate that establishments which have relocated production capacities abroad within the last two years are far less flexible in reacting to customers' orders and at the same time the time to market is slowed down substantially. In the increasingly competitive "race against time," this time lag may prove to be the decisive negative factor for the market success of an enterprise.

In view of this, the increasing tendency of small and medium-sized establishments towards future relocations of production must be regarded critically. The perception by these establishments that they must have a presence abroad, increases the danger of relocation decisions being made in a hurry. Thus, the possible negative consequences for flexibility and innovative capability may often be left out of consideration when reaching a decision, despite the fact that good reactive ability and flexibility are central competitive advantages that need to be maintained for small and mid-sized manufacturers. Neglect of parameters such as these may result in painful costs in the medium term, which may make it apparent after the event that relocation decisions were a step in the wrong direction. However, these decisions are very hard to reverse, as they usually involve a loss of personnel-related know-how. The threatened loss of core competencies may bring great problems, particularly for small and mid-sized firms with their limited resources.

On the other hand, relocations for reasons of cost do appear to raise the average productivity of locations in Germany. However, the unproductive production processes are merely being "rationalized" out of existence. The parts of production that remain in Germany are not necessarily improved. It is suggested that an alternative to this situation is the use of new production concepts. A consistent reorganization of the processes taking place at the original location in Germany may well improve productivity to such an extent – without negatively affecting flexibility, innovative capability or core competencies – as to exceed the advantages that might be achieved by a foreign relocation. This option would even appear superior to a combined strategy of relocation plus restructuring of remaining areas of production.

Only large-scale producers of simple products achieve less productivity by modernizing their production structures than by relocating production capacities. However, in this group of establishments, the new production concepts seem to be used mainly to improve performance parameters (rather than to improve productivity). This demonstrates that the potentials of the concepts are definitely suited to the pursuit of different business strategies. On the one hand, their rationalization potentials may be the central aim of reorganization measures supporting a cost reduction strategy. Producers of complex goods exploit these cost reduction potentials to the full, but are not able to improve other performance parameters such as flexibility at the same time. New production concepts can be used, on the other hand, to compensate the "locational" disadvantages of German enterprises by strengthening performance-oriented competition factors. As demonstrated by the example of the manufacturers of simple products, it is then only possible to increase productivity to a limited extent. Thus the scope for distributing the effects of new production concepts is limited (see chapter 4).

In the light of these considerations, two alternative options are open to German companies when making a cost-oriented decision on relocation. First, firms can apply the new production concepts to improve performance-oriented parameters apart from the product price, like flexibility or time to market, in order to compensate productivity disadvantages. Second, they can try, by consistent reorganization, to optimize the productivity of their processes within Germany to such an extent that the cost reduction potentials of a possible relocation are more than compensated. Unlike relocation, both these alternatives mean that the whole of production remains in Germany, thus contributing to securing jobs at home.

8.7 Bibliography

Kinkel, S. (1996), Wer Produktion ins Ausland verlagert, verschenkt Verbesserungspotentiale im Inland: Flexibilität und Produktivität sind angesichts des Arbeitsplatzabbaus bei Produktionsverlagerungen kritisch zu hinterfragen, PI-Mitteilungen, No. 2, Fraunhofer Institute for Systems and Innovation Research, Karlsruhe.

Lay, G. (1997), Prozeßinnovationen als Schlüssel zu innovativen Produkten: Mit neuen Produktionskonzepten in Wachstumsmärkte, PI-Mitteilungen, No. 7, Fraunhofer Institute for Systems and Innovation Research, Karlsruhe.

Lay, G., and Mies, C. (Eds.) (1997), *Erfolgreich reorganisieren. Unternehmenskonzepte aus der Praxis*, Springer, Berlin, Heidelberg 1997.

Lay, G., Dreher, C., and Kinkel, S. (1996), Neue Produktionskonzepte leisten einen Beitrag zur Sicherung des Standortes Deutschland, PI-Mitteilungen, No. 1, Fraunhofer Institute for Systems and Innovation Research, Karlsruhe.

9 New Production Concepts and Service Orientation: The Case of Teleservice

Carsten Dreher, Gunter Lay, and Thomas Michler

9.1 Introduction

Goods made in Germany can not compete primarily through low prices. German production input costs, particularly the expense of labor, are too high for that. Over the last few decades, German manufacturers have successfully compensated for their disadvantages in production costs by focusing on advanced technology, high quality, and specialized products tailored to customer needs. However, recently an additional strategy has been added to this portfolio – to package together manufactured goods and complementary services (Reichwald and Möslein 1995, Ganz and Stanke 1996, Homburg and Garbe 1996). The basic idea is that the complementary services add value to the goods and distinguish them in the market, raising the package above the intense global price pressures that would otherwise apply (Simon 1993). If enterprises succeed in melding goods and services into a combination which customers can order like a service, disadvantages based solely on manufacturing costs do not become so relevant.

In the investment goods sector, there are several ways in which goods and services can be integrated. Traditionally, many producers of machines offer maintenance and repair services as well as training for customer employees. New approaches have also been introduced through which products are leased rather than sold to customers. These approaches allow vendors to offer regular upgrades to adapt machines to new technologies and standards or to run the machines with own personnel rather than the customers' (see Lay 1996).

The increasing complexity of machinery and equipment is one of the reasons that make the packaging of services and goods attractive to customers. The diagnosis and correction of faults places ever-higher demands on service personnel in terms of skills and qualifications. In their search to reduce costs, manufacturers use their expensive machines intensely and want high availability. The manufacturing process has also become increasingly integrated, as buffer inventories have been slashed, individual manufacturing steps tightly linked to one another, and delivery times shortened. Long down times for machine repairs thus cause not only costly but also intolerable delays.

For several years, some investment goods manufacturers have supplemented their product range by offering "teleservice" to rapidly supply specialized know-how to maintain and repair equipment (see Malle 1996). Teleservice usually involves the automatic recording of data from machines and processes operating in a customer's factory, the analysis of this data in diagnostic program, and the transfer of the information through an electronic network to the machine supplier's service center (Adam, Bistram and Linnemann 1996). In this way, it is possible for specialists to identify problems – and initiate measures for correcting them – remotely. Response time is quicker and travel is avoided, thus reducing delays and costs.

This chapter looks at the extent to which German industry make use of these systems to monitor machines and plants via teleservice, drawing on the 1995 ISI manufacturing innovation survey of the German investment goods sector. Patterns of teleservice use by establishment size and industry are examined. Whether teleservice replaces the users' technical staff is also examined. These analyses provide a basis not only to examine the direction and pace of teleservice expansion in the future, but also to assess implications for business strategy and competitiveness for suppliers and users.

9.2   The Diffusion of Teleservice

The use of teleservice in German industry has developed gradually over the last twenty years. Although the first pilot applications of teleservice got underway in the second half of the 1970s, it was not until the late 1980s that diffusion picked up. Since the beginning of the 1990s, increasing numbers of firms have adopted teleservice. By 1995, about one quarter (24 percent) of plants in the investment goods industry in Germany were making use of the option of having plants and machines monitored by teleservice (figure 9-1).

Teleservice is now diffusing at a growth rate that is rising annually. In theory, it might be expected that teleservice will follow other innovations, by diffusing along an "S-shaped" pattern. Thus, further increases in diffusion are possible, before the diffusion rate slows. Moreover, if one looks at the extent or density of use *within* establishments, as opposed to the aggregate user rate, the potential for growth may be even higher. On average, it seems that users only apply teleservice to about one-half of the possible applications in their facilities. Thus, in the future, teleservice

Figure 9-1   Diffusion of teleservice in the German investment goods sector

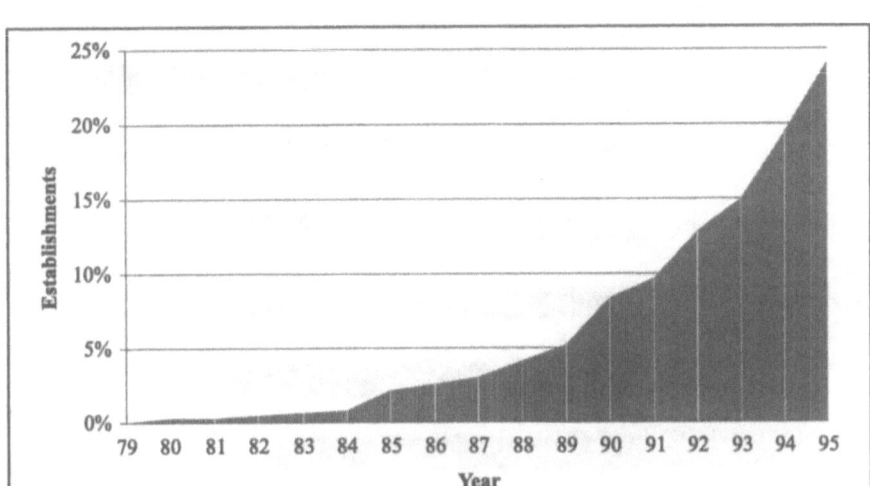

could be of great interest not only for the three-quarters of firms that are still non-users, but also for firms that have already implemented this innovation. However, although the *potential* may be there for significant expansion in teleservice use, at present most of these non-users have not yet developed *concrete* plans to invest in teleservice, as a later section of this chapter discusses.

9.3   Teleservice Use by Industry and Establishment Size

Within the broad aggregated data on teleservice diffusion, there are important variations in use among different sub-sectors of the German investment goods industry. The greatest diffusion is found in the mechanical engineering industry, where 28 percent of establishments use teleservice. Among producers of automobiles, electrical engineering products and precision instruments and optical equipment, slightly over one-fifth of establishments are teleservice users. Lower teleservice diffusion is evident among producers of sheet metal goods and structural metal products, with user rates of 19 and 13 percent respectively (figure 9-2).

However, there is also a wide range of user intensities within the mechanical engineering industry itself (figure 9-3). Textile machinery producers have the highest teleservice usage rates, at 41 percent, while among agricultural machinery producers teleservice is used by only 14 percent of establishments. Since (as other studies have shown) textile and clothing manufacturers are themselves frequent teleservice

Figure 9-2   Teleservice use by industry, investment goods sector

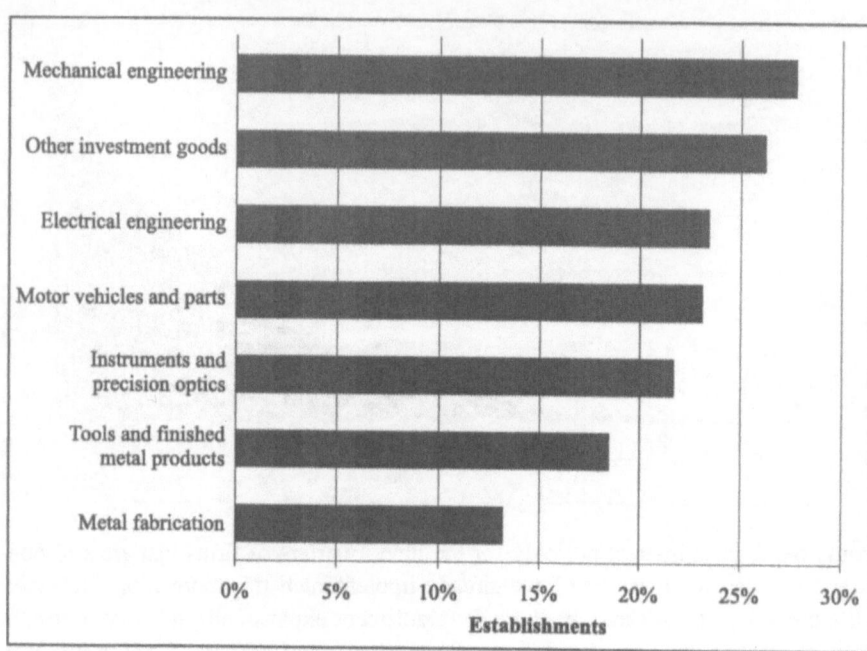

users, this may be an instance where high customer-demand for teleservice encourages the suppliers of textile machinery equipment to also adopt teleservice in their own production.

Another important structural characteristic affecting the diffusion pattern of innovation is the size of the user. Not surprisingly, our analysis shows the type of correlation typical for the use of innovative production technologies, with larger firms implementing teleservice more often than smaller ones. As many as 39 percent of establishments with more than 500 employees use teleservice solutions. The user rate for plants with 100 to 500 employees is 27 percent. Only 20 percent of small manufacturers with less than 100 employees make use of teleservice (figure 9-4).

The idea that teleservice might substitute for in-plant maintenance personnel is not supported by our data. Small plants are less likely than large plants to have resident maintenance specialists and thus might be expected to find teleservice an effective option to more quickly resolve maintenance and repair problems. But it is larger plants which are more likely to have resident specialists to service and repair their equipment that are increasingly taking advantage of teleservice to complement their own capabilities, as section 9.4 discusses.

Figure 9-3   Teleservice use within the mechanical engineering industry

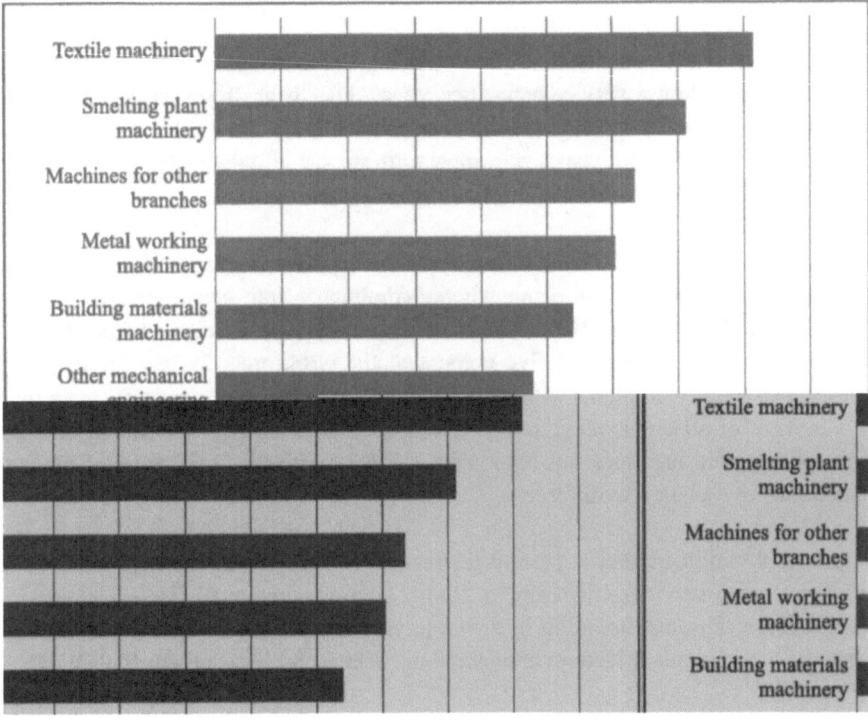

Figure 9-4   Teleservice use by establishment size, investment goods sector

9.4   Teleservice: An Alternative to an Internal Service Department?

We conducted further analysis to probe whether teleservice has been adopted as a *substitute* for in-house service capacities, or whether it has been used primarily to *complement* internal servicing and maintenance facilities. To do this, we correlated the employment of maintenance personnel with the use of teleservice. However, we found no relationship between these two variables.

On average, four percent of the employees in the investment goods industry are engaged in maintenance. For plants where maintenance staff constitutes two percent or less of total personnel, 24 percent are using teleservice. In plants where maintenance staff makes up three to five percent of the workforce, 25 percent are using teleservice. Finally, in plants with a high proportion of maintenance staff (more than five percent of all employees), teleservice is used by 23 percent of establishments. Thus, teleservice use does not obviously correlate with the percentage of maintenance staff in a plant's workforce.

These results indicate that at present teleservice is not regarded as an alternative to the firm's own servicing department. Rather, the advantages of teleservice are regarded more as complementing in-house service facilities, with teleservice support from the machine manufacturer enhancing the users' problem-solving capacities.

9.5   The Future Development of Teleservice in German Industry

Against this background, how will the use of teleservice develop in the future in Germany? Where are the potentials for increased use of teleservice in the investment goods industry to ensure maximum availability of machinery and factories?

Of the establishments that are already using teleservice today, one quarter are planning to further extend their use of this innovation in coming years. This expansion might mean installing telemonitoring for more machines, or making qualitative improvements to existing teleservice options. The planned extensions signal that these respondents have had positive experiences in using teleservice and are willing to make further investments in the future. Of the firms that have not used teleservice so far, 13 percent have concrete plans for adopting this innovation, while 87 percent do not (yet) consider its use to make sense for them. Thus, the firms now planning to become first-time users of teleservice represent a 10 percent share of the whole investment goods sector.

Figure 9-5   Future teleservice plans in the investment goods sector

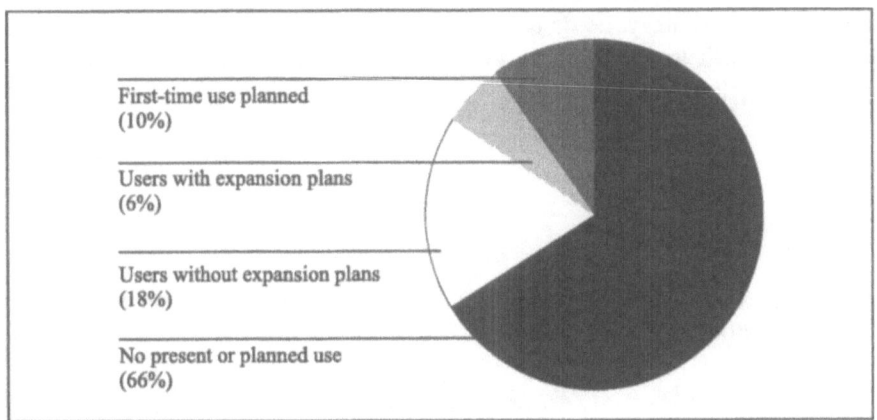

First-time use planned
(10%)

Users with expansion plans
(6%)

Users without expansion plans
(18%)

No present or planned use
(66%)

If the plans by users and by plants that have not previously made use of teleservice are combined, it can be seen that about 16 percent of manufacturers in the investment goods industry expect to start new or expanded teleservice projects. This implies that the diffusion rate will go up from one quarter to one third of firms. This growth stands in contrast to the largest group of firms – representing two thirds of the investment goods industry – who do not at present consider teleservice to be an appropriate way of providing support for their production.

9.6   Conclusions

Germany is said to have an international lead in terms of the range of teleservice options available and the technical level of solutions currently offered by suppliers (Hudetz and Harnischfeger 1997). Indeed, teleservice has become an innovation that has gained broad acceptance in German industry in assisting the smooth running of manufacturing production processes. This is illustrated by our data on diffusion in the investment goods sector, with 24 percent of plants who are users today, and growth to up to 33 percent of users expected over the next few years. Teleservice appears to complement existing in-house capabilities, rather than substitute for them.

However, the majority of plants that currently are nonusers still have no specific plans to introduce teleservice. It is possible that the companies that remain skeptical about teleservice are concerned about incompatibilities with their existing equipment and procedures. This is a typical reaction among non-users of new industrial

technologies. Proof that teleservice can fit with firms' own servicing and maintenance concepts, and that machine suppliers can support teleservice as part of an
overall service strategy, needs to be demonstrated to non-users for further significant market penetration to occur. Efforts to overcome this hurdle will open new
diffusion possibilities.

9.7   Bibliography

Adam, W., and Bistram, R., Linnemann, H. (1996), Fernbetreuung von CNC-
gesteuerten Fertigungsanlagen, *VDI-Z*, 3, pp. 84-87.

Ganz, W. and Stanke, A. (1996), Design hybrider Produkte. In: Volkholz, V., and
Schrick, G., (Eds.), *Dienstleistungen im 21. Jahrhundert*, RKW-Verlag,
Eschborn, pp. 85-92.

Homburg, C. and Garbe, B. (1996), Industrielle Dienstleistung als Manage-
mentherausforderung, *Management Zeitschrift*, 65, pp. 31-35.

Hudetz, W. and Harnischfeger, M. (1997), *Teleservice für die industrielle Produktion - Potentiale und Umsetzungshilfen*, Forschungszentrum Karlsruhe FZK-
PFT 186, Karlsruhe.

Lay, G. (1996), Integration von Produktherstellung und Dienstleistung als Strategie
zur Sicherung der Wettbewerbsfähigkeit in einer globalisierten Wirtschaft. In:
Hoss, D. and Schrick, G., (Eds.), *Wie rational ist Rationalisierung heute?*
Raabe-Verlag, Stuttgart, pp.57-67.

Malle, K. (1996), Prozeßkette Blech - Technologie und Teleservice, *VDI-Z*, 5, pp.
12-13.

Reichwald, R. and Möslein, K. (1995), *Wertschöpfung und Produktivität von
Dienstleistungen? - Innovationsstrategien für die Standortsicherung*,
Arbeitsberichte des Lehrstuhls für Allgemeine und Industrielle
Betriebswirtschaftslehre, Band 6, TU München.

Simon, H. (1993), Industrielle Dienstleistungen und Wettbewerbsstrategie. In:
Simon, H. (Ed.), *Industrielle Dienstleistungen*, Schäffer-Poeschel Verlag,
Stuttgart, pp. 3-22.

10   Environmental Practices in Germany's Investment Goods Sector

Jürgen Fleig

10.1   Environmentally Clean Production in German Industry?

Environmental conservation has an important position in public discussion in Germany and has great significant in German politics, legislation, and regulation. Environmental laws and state funding incentives induce – or force – businesses in Germany, to pay attention to environmental conservation in manufacturing. In addition, environmental conservation in industry is also promoted by voluntary codes of compliance.

The external pressures require German enterprises to invest for environmental conservation. Since at least some of these environmental investments are necessary to meet specific regulations that vary by industry and process, there are great differences in environmental investments between different industries. This is evident if one compares environmental spending for the investment goods sector with other branches of German industry, measured by the share of environmental investment in all investment (figure 10-1). Although investment goods firms spend a lower share of their total investments on environmental protection, business advocates are concerned that expenditures on environmental conservation technologies represent an additional cost-factor that endangers the competitiveness of the German industry. On the other hand, environmentalists argue that conservation measures not only lead to a cleaner environment but also save money through waste reduction and better efficiency as well as open-up new markets for clean products and technologies.

There has been a long-run increase in the share of investment allocated for environmental protection in the German investments goods sector, from 1.6 percent in 1979 to 2.8 percent in 1994 (figure 10-1). The highest levels of environmental investment (more than 3 percent of the total) were seen in the first half of the 1990s. This resulted largely from higher investments in technologies to reduce the cost of

114

Figure 10-1Figure 10-1   Investment share for environmental conservation in German industry, 1979-1994

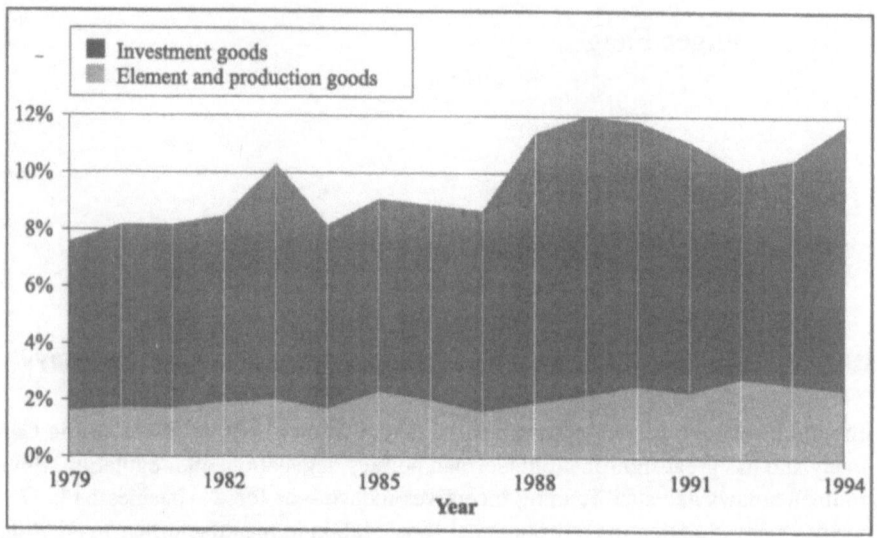

Source: German Statistical Yearbook.

sewage and waste disposal. But, overall, environmental conservation investments are not very high in the investment goods sector, and have declined slightly as a proportion of all investments in recent years. Moreover, between 1988 to 1994, only one in eight firms in the investment goods sector made any environmental investments at all. At least for the investment goods sector, the expense of environmental protection is perhaps not as high as the frequency of public discussion might imply.

To probe more fully the practices and implications of environmental expenditures in the German investment goods sector, this chapter examines two operational procedures. The first is more technical – the use of "dry processing" in manufacturing production. The second is more organization – the use of environmental audits. The basis for analysis is the Fraunhofer ISI manufacturing innovation survey (see Kinkel and Dreher 1996, and also the appendix to this book).

Dry processing involves manufacturing production steps like twirling, molding or drilling that are undertaken without cooling lubricants. This can be achieved by material substitution (using aluminum or plastic) or by changes in manufacturing procedure that are undertaken with help of suitable machines, so that no corresponding lubricants are required.

In some cases, it is very easy to work without cooling lubricants; no other changes within the process are required besides cutting off the inflow. Therefore one can find dry processes in use for thirty or forty years or even earlier, since the initial development of machine-based twirling, molding or drilling. In most other cases, changes that are more significant must be undertaken, especially in the materials used for cutting tools, the design of machines, and the quality of the material. The cost of disposing of used cooling lubricants in compliance with environment regulations has risen since the end of the 1970s and early 1980s. This has prompted innovations to reduce or to avoid cooling lubricants: new materials were developed and machines were redesigned, among other changes. However, some difficulties remain. Even with the new technology, it is often not easy to guarantee the necessary quality of final products. High fixed costs and overheads in changing materials, purchasing machines or training can only be reduced if all steps in manufacturing are changed. But this is not always possible.

Procedures for environmental auditing involve organizational as well as management actions to assess whether an enterprise is complying with its own or external standards to protect the environment. Environmental audits are necessary to meet some regulations. The most important environmental audits are the "Environmental Management Audit System" or EMAS (Öko-Audit 1993) and the ISO 14000 international standard. In both cases the environment-management-system of an enterprise is externally audited and, if in compliance, certified.

The EMAS consists of several steps to achieve certification. First, the company must develop its environmental policy. Second, it conducts an initial environmental audit to identify weaknesses and potentials. Then it formulates an environmental program. From this program, environmental management goals are derived, which are very important because the company will be assessed on its achievement of these goals and its continuous improvement on the underlying criteria. In parallel the company develops a management system involving the allocation of responsibilities and the development of tasks and measures. The company should also develop a handbook or checklists for all employees. Information about this environmental system should be included in a publicly accessible company environmental statement. An external referee audits the management system and the statement. If successful, the company receives a certification that is recognized by the European Commission. Finally, there will be further internal audits to see whether the measures brought some improvements and if the management system works. ISO 1400 also requires extensive internal company documentation of environmental management procedures, an external audit, and certification to standards. Despite these similarities, environmentalists say that there are some fundamental differences, pointing out that, compared with EMAS, ISO 1400 does not require as great a company commitment to environmental protection and improvement and there is no requirement under ISO 14000 to make a public environmental management statement.

Dry processing and environmental auditing (through methods such as EMAS and ISO 1400) each represent measures through which companies can address demands for environmental compliance. Both measures are voluntary; in general, there are reasons besides legal necessity for businesses to implement these measures. Businesses are thus likely to undertake the measures (each of which involves expense) only if they see some corresponding benefits. Some businesses may implement these measures proactively to earn forerunner status and to improve their public image. Others will hope for economic returns, particularly through reductions in waste and improved efficiency. Since both measures are voluntary, it might be expected that companies would only implement them if they are economically successful and have a favorable profit situation. This particular hypothesis is explored in detail after an analysis of the diffusion rates of the two environmental techniques.

10.2   Dissemination of Dry Processing Technologies

The share of the manufacturers in Germany's investment goods industry that have implemented dry processing is about 17 percent, according the 1995 ISI survey. The diffusion rate is slightly highly in West Germany, at 18 percent, than in East Germany with 16 percent (figure 10-2). In 1989, there was a gap between the two regions of 6 percent, but after German unification, East German manufacturers rapidly upgraded their technological levels. Investments were made in machines (with government support) to revamp equipment, bringing in modern technologies like dry processing.

The trends for the diffusion of dry processing suggest that it will reach a saturation level, under current conditions, of about 20 percent; in coming years, only 4 percent of the manufacturers plan to introduce the technology for the first time. This implies that the remaining 80 percent of manufacturers see no need or possibility to adopt dry processing. About one fifth of these companies have a very low share of mechanical production (under 5 percent) in their plants; for these the adoption of dry processing is probably not practical. But for the majority of companies, dry processing is not of interest because of technical or economic reasons. Possibly the necessary requirements for quality can not be guaranteed by this technology, or the technology is not cost-effective given numbers or range of products items produced.

Figure 10-2   Dissemination of dry processing in the German investment goods
                sector

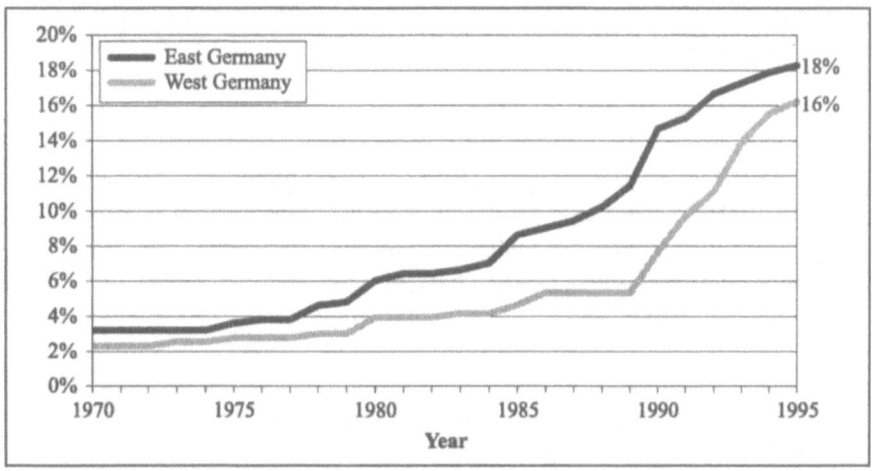

Some companies may only use dry processing in parts of their manufacturing,
meaning that there may be potential for greater deployment even among existing
users. Respondents to the survey were asked to make their own assessment of the
proportion of the potential for dry processing that they actually exploited. But this
is hard to define: it can be interpreted to mean the workload of the machines, the
part of the production program that uses dry processing, or the performance of the
process. However, on average, respondents say that they exploit only about 50 percent of the technical potential of this environmental protection technology. By
comparison, CNC technology is exploited on average to about 70 percent of its
potential. There is also a wide range in the levels of exploitation of dry processing
within individual manufacturers. One quarter of respondents say they exploit less
than 20 percent of the possibilities for using dry processing, while another report an
exploitation level of 85 percent. This span suggests that the control and full exploitation of the technology requires a learning process for many companies whose
success depends on numerous influential factors.

A more detailed analysis of the companies also suggests that using dry processing
is only meaningful under certain conditions. Companies making customized client-
specific products use dry processing less frequently, with only 19 percent planning
further use of the technology. In contrast, 26 percent of the plants making standardized products plan an extension of its capacities of dry processing.

Larger plants use dry processing more frequently than small ones. More than 40
percent of large plants (with more than 1000 employees) use this technology –

compared with only about 10 percent of smaller establishments (1-19 employees). Among other factors, the reason for this may be that the investment required for dry processing is too big for small manufacturers, or it that it is not worthwhile if only a small number of pieces are produced. Large plants have more possibilities to test a new technology like dry processing (in a part of their factory) – and, if necessary, to develop it further.

There are also differences in the diffusion of the dry processing by industry. The automotive industry is at the forefront, with 27 percent of establishments using dry processing. In mechanical engineering industry, the diffusion rate is 22 percent. Adoption rates are much lower electrical engineering with approximately 5 percent. In this branch, machining and molding is considerably less important than in the automotive or mechanical engineering industry.

10.3   Dissemination of Environmental Audits

Traditionally, technological approaches to environmental conservation (such as dry processing) have played a much bigger role than organizational methods. Only in the last few years have businesses recognized that organizational measures can also help to reduce of materials and energy consumption and assist in achieving environmental compliance. But progress in introducing organizational measures has been slow. By mid-1997, only about 700 firms have been certified through the Environmental Management Audit System or EMAS (DIHT 1997). Among investment goods manufacturers in Germany, just 6 percent have conducted their own or a standardized Environmental Audit, with the share clearly higher in West Germany than in East Germany.

However, environmental audits will play a bigger role in the future. More than a quarter of the manufacturers that have not conducted an environmental audit up to now plan to do so in coming years. Different reasons underwrite these plans. Some manufacturers are under pressure from major customers to implement environmental auditing and certification. For other firms, environmental audits promise to reduce personal management risks due to environmental regulations and liability. Finally many companies seek an environmental audit to maintain or improve their ecological image to the public.

Large manufacturers are in the lead with respect to environmental auditing (figure 10-3), although absolute levels are not high. Barely 30 percent of establishments with more than 1000 employees have implemented an environmental audit. About 8 percent of mid-sized plants with 200 to 500 employees use environmental audits,

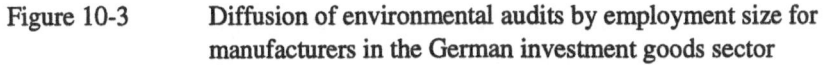

Figure 10-3   Diffusion of environmental audits by employment size for
                manufacturers in the German investment goods sector

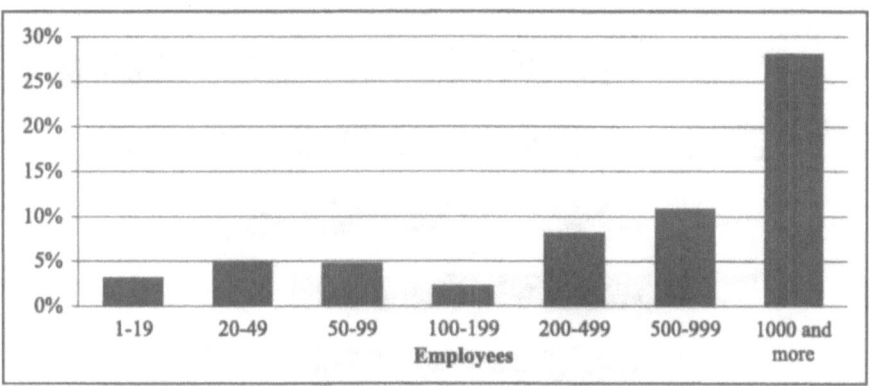

while less than 4 percent of plants with under 100 employees use this organizational approach to environmental protection.

Thus, up to now it seems that it is mainly larger plants who can or want to afford an environmental audit. Many examples show that the expenditures connected with such auditing and the necessary know-how are not insignificant. Additionally, the necessary continuous improvements involved in an environmental audit need additional staff and expenditures.

There are also some differences in the diffusion of environmental audits by line of business (figure 10.4). The automotive industry is the most frequent user, followed by steel fabricators and manufacturers of iron and sheet metal goods and metal hardware. This illustrates that requirements for environmental conservation are different in particular industries. Some industries, such as automotive production, are more strongly in the state and public view (for example, by trade supervision offices or potential customers); they must do more for the environmental protection and be proactive.

10.4   Conditions of Application and Effects

Environmental protection technologies (for example, dry processing) or organizational measures (such as environmental audits) require additional investments and expenses. Some manufacturers nevertheless take such measures in order to fulfill

Figure 10-4   Dissemination of environmental audits by industry, German investment goods sector

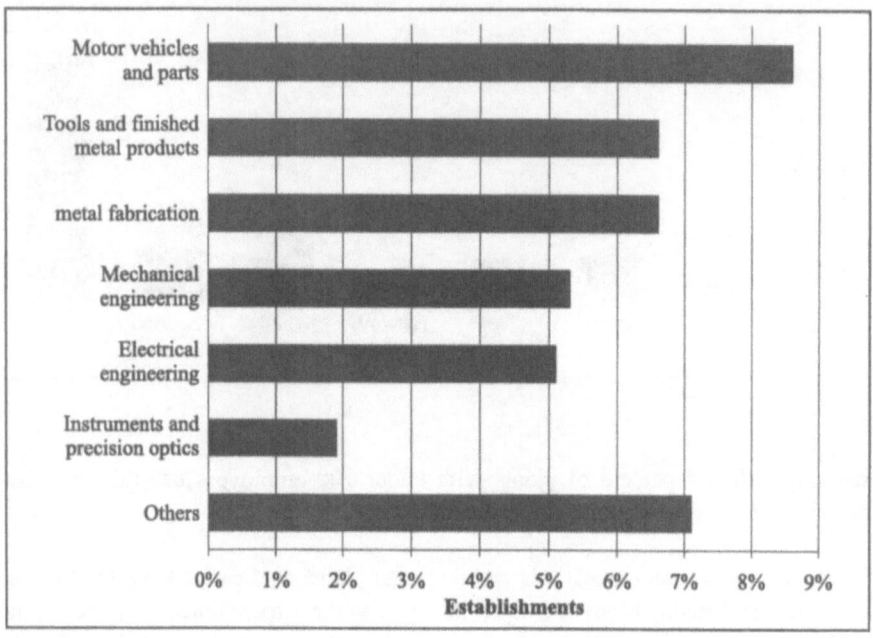

the requests of the customers, to improve their image or to reduce the risk to management. These effects represent no immediate advantages in competition, although there may be indirect benefits. Therefore, it is plausible to suggest that manufacturers only deploy such voluntary measures for environmental conservation in economically good times – when they are profitable and have available resources to cover the expenses involved.

To check this hypothesis, an assessment of manufacturers' self-reported profitability was examined. This indicator expresses directly how manufacturers see their economic situation. We can thus see whether firms that have implemented environmental conservation measures are more likely to have good or satisfactory profit situations.

For users of dry processing technologies, there is no difference in this assessment of profitability (figure 10-5). Here, the thesis cannot be confirmed. However, for environmental audits, the comparison looks somewhat different. For manufacturers that use environmental audits, none has a profit situation that is acutely dangerous (compared with 3 percent of non-users whose profit situation is acutely dangerous). About 28 percent of manufacturers with environmental audits have an unsatisfac-

Figure 10-5   Present and expected profit-situation of the businesses and use
of environmental audits

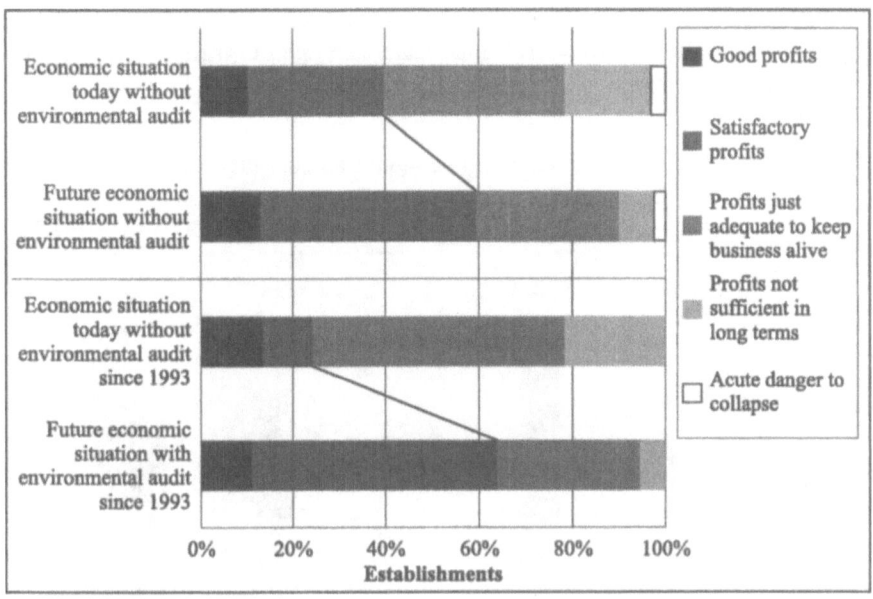

tory profit situation – about 10 percentage points higher than manufacturers without environmental audit. On average, more companies with poor profit situations use environmental audits. In other words, this measure for environmental conservation is not used only by companies with a good profit situation.

Rather, it seems as if manufacturers use an environmental audit in order to improve their outlooks for the nearer future. Plants that show a bad profit situation at present have the opinion that an environmental audit will contribute to an improvement in the future. Plants that have enforced an environmental audit in the last three years look essentially more optimistically into the future, although their profit situation is worse today, than companies without environmental audits.

If manufacturers proactively pursue measures for environmental conservation, this may also be related to their potential for innovations and their innovative self-image. The share of the companies that have developed and marketed new products in the last three years can be an indicator for this; such companies can be called innovative in general. Maybe these innovative companies are also more likely to be progressive in other areas, such as environmental protection.

Indeed, the survey results show than manufacturers who have adopted dry processing technologies or an environmental audit are more innovative than non-adopters (figure 10.6). The share of the plants with new products and with environmental audits is about two percent points higher, with dry processing eight percentage points highly. These are not huge differences. Nevertheless it shows that companies with an innovative self-image are also open to measures for the environmental conservation.

Figure 10-6   Effects of environment-audits and dry processing technologies on economic features

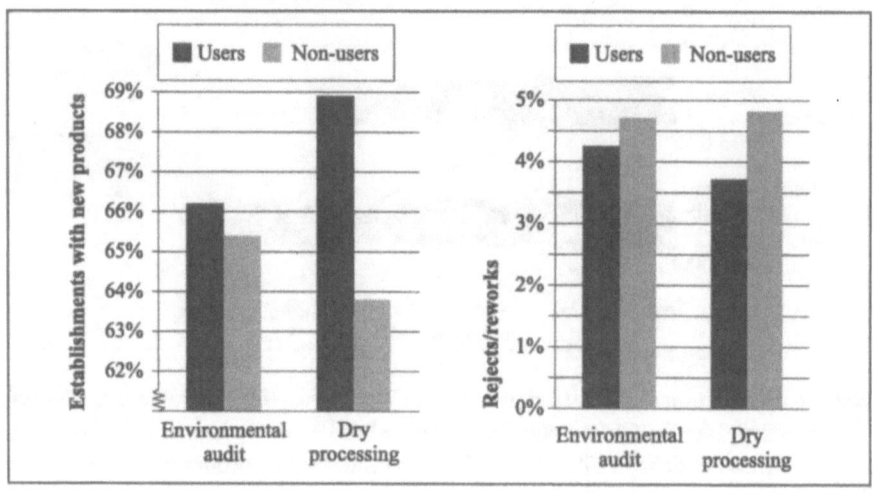

In addition to profit-situation and innovation-ability, a third aspect that might be related to use of environmental measures is the management of production in the factory. An indicator for this capability is the quota of rejects or necessary rework (average share of all produced products). Plants with low reject rates manage their processes better in general and can guarantee a high quality level. It would be plausible to suggest that these companies are more likely to adopt new technical and organizational measures. Indeed companies with environmental audits or with dry processing technologies have lower rates of rejects than non-users (figure 10.6). With dry processing, this is very clear (although the statistical test indicates only a very small significance at the 0.12 level): 3.7 percent versus 4.8 percent rejects, or a 23 difference. With the environmental audit, the differences are smaller: 4.3 percent versus 4.7 percent rejects, with which this difference could also be influenced by the different diffusion rates by particular industries. Still, this does suggest that companies that manage their production processes with high quality are also open for new measures for the environmental protection.

10.5   Conclusions

The analysis of the survey results shows that technical and organizational environmental conservation still plays a relatively small role in the German investment goods sector. Direct environmental investments as a share of all investments are low, even if further cost-effective measures follow which are not captured in the official statistics. Based on this data, it is hard to argue that environmental investments are hampering the economic capabilities of these companies. Moreover, the diffusion of the dry processing and of environmental audits is not very high, particularly in the comparison to other new technologies (for example teleservice as discussed in chapter 9).

Nevertheless a small part of sector is taking proactive measures to protect the environment. Innovative plants are most likely to experiment with technical procedures such as dry processing. Indeed this procedure has only restricted possibilities of further adoption in its present form and is unlikely to diffuse much more in the near future. Environmental auditing has gains attention in recent years. But, in view of the high cost of a certified audit, large companies are most able to afford this at present.

However, one hopeful point is that measures to promote environmental conservation are not dependent on the profitability of the company is positive – they are not "measures for good times only", that is not heeded under adverse economic conditions. Indeed an opposite trend is found. Manufacturers that assess their profit-situation as less good than others seem to see the environmental auditing as a chance to improve their profit-situation. In addition, the adoption of environmental conservation measures is promoted by the innovative self-image of a company; innovative companies are more frequently active in the environmental conservation. Manufacturers who manage their internal processes with high quality are also comparatively more open to new technical and organizational measures for environmental conservation.

10.6   Bibliography

DIHT (1997), *Umweltmanagement – Standorte in Deutschland*, Deutscher Industrie- und Handelstag, <http://www.diht.de/>.

Kinkel, S. and Dreher, M. (1995*), Produktionsstrukturen der Investitionsgüterindustrie Deutschlands*, Methodische Grundlagen zur Auswertung der Umfrage, Arbeitspapier, Fraunhofer Institute for Systems and Innovation Research, Karlsruhe.

Öko-Audit-Verordnung (1993), Verordnung (EWG) No. 1836/93 des Rates vom 29. Juni 1993 über die freiwillige Beteiligung gewerblicher Unternehmen an einem Gemeinschaftssystem für das Umweltmanagement und die Umweltbetriebsprüfung. *Amtsblatt der Europäischen Gemeinschaften*, 107, No. L 168/1-18.

Statistisches Bundesamt, (Eds). (1985 and succeeding years), *Investitionen für Umweltschutz im Produzierenden Gewerbe*, Fachserie 19, Metzler-Poeschel, Stuttgart.

11   Adoption of New Production Concepts in East Germany

Gunter Lay

11.1   Introduction

After the economic and monetary union between East and West Germany in 1990, industrial firms in the former German Democratic Republic (GDR) were abruptly exposed to the production and trade requirements of a market economy. These new requirements were radically different from the business conditions that had been established under economic planning. In addition to changes in markets, ownership, and control, industrial enterprises in the former GDR had to fundamentally restructure manufacturing strategies and practices. It was necessary to improve marketing as a prerequisite for selling their output under the new conditions and to increase the intensity of research and development as a precondition for making competitive products. In addition, these enterprises had to modernize facilities, equipment and operations as a means for competitive manufacturing of products.

The modernization of production was to be achieved predominantly by investment measures aimed at increasing productivity. Focusing this investment on the acquisition of advanced technologies appeared to be the most appropriate method for firms in the new federal Länder (states) of the former GDR to match national and international productivity standards (DIW 1995, p. 535, Felder et al. 1995, p. 10, Dietrich 1997, p. 5).

Many enterprises in the former GDR pursued this strategy of technological modernization – aided by public incentives and subsidies (Lay, Michler and Pleschak 1995, p. 55). Indeed, just a few years after German economic and monetary union, there was some evidence that the standard of equipment and machinery installed in factories in the new Länder was rising to the technological level found in western enterprises (Wengel and Harnischfeger 1995, p. 97, Ostendorf 1993, p. 21).

However, despite these investments in new equipment and machinery, the economic performance and productivity of enterprises in the new federal Länder has improved much more slowly than was expected (DIW 1997, p. 45, Gürtler 1997, p. 10). One indicator for this can be seen in the level of productivity. In 1996, the gross productivity per employee in the manufacturing sector in the new Länder was only 55 percent of the West German level according to federal statistics (Ragnitz 1997b, p. 3). Poor productivity is one of the factors that have contributed to the continuing economic and competitive weakness of the manufacturing sector in East Germany.

This chapter probes the connection between the modernization of production facilities and productivity change in East Germany. In so doing, it seeks to systematically examine why new technological investments have not resulted in the anticipated operational and economic outcomes. The chapter starts with a review of the use of modern production technologies in East Germany at the point of economic and monetary union. This is followed by an assessment of the period following German reunification, to consider the extent to which enterprises in the east subsequently caught up with the constantly evolving standards of production technology applied in the west. The chapter then analyzes the productivity impacts of different modernization strategies in East German enterprises. Finally, the chapter examines the importance of new production technologies compared to other factors in influencing the productivity of East German manufacturers.

The chapter draws on the 1995 ISI manufacturing innovation survey. Of the 1,305 industrial respondents to this survey, 558 came from firms in the new federal Länder and 747 from enterprises in the old federal Länder (see appendix for details on the survey methodology and response rates). One of the survey's major objectives was to develop data to evaluate government efforts to promote the use of new computer-integrated manufacturing technologies in the new federal Länder. By design, this led to an over-sampling of East German enterprises. There were also other differences in industrial characteristics of respondents – in particular, an over-representation of small East German establishments and an over-representation of West German machinery manufacturers. The size and industry of firms has an influence on the use of production technology solutions (Lay and Michler 1990, p. 79). Smaller firms are less likely to use new technology, while industry sectors such as machinery producers are among the more advanced. To adjust for these variations, a weighting scheme was used. The weighting scheme for the data reported in this chapter compensates for sampling differences between east and west Germany and ensures that there is a comparable industrial and size structure. This enables a comparative analysis of technology diffusion between the two regions. However, if the weighted East German data is considered in isolation it is possible that East German technology use will be overestimated.

11.2   The Modernization of Production in the New Länder

This section compares the adoption of new production technologies in the new and old Länder. Four technologies are examined (see figure 11-1):

- *Computer aided design* (CAD) – a technology that supports and enhances proficiency in the product design and development process. CAD is often regarded as a prerequisite to acquire orders from certain groups of customer or to quickly meet special customer requirements.

- *Product planning and control systems* (PPC) – a technique to increase the efficiency of materials management. PPC aims to decrease inventories, thus saving capital, and to improve the flexibility of materials supply in production.

- *Computer numerical control (CNC)* of machine tools - a technology for the flexible automation of parts manufacturing. CNC enables firms to manufacture small batches more productively than with conventional machine tools. At the same time, CNC machine tools offer the option of shifting from large-batch production runs made on dedicated machines to customized medium-sized production runs. This change can improve flexibility and maintain productivity.

- *Computer-integrated manufacturing* (CIM) – the integration of computer-controlled design and manufacturing technologies, including the integration of CAD and numerical control (NC) technologies. CIM generated high expectations of improved production effectiveness, although the mixed experience of enterprises in the old Länder during the nineteen eighties with CIM should have signaled that these expectations would not be easily realized.

Among investment goods manufacturers in the new Länder, CAD was hardly in use before the economic and monetary union. In 1988, only about 7 percent of GDR establishments used this technology. At the same time, however, nearly a third of the establishments in the old Länder were already using CAD - a difference in diffusion rates of 23 percentage points. During the first half of the nineties, the diffusion of CAD spread rapidly in western firms – 79 percent were using CAD systems by 1995. Nonetheless, plants in the new states exceeded even this rapid growth rate. CAD use jumped from 7 percent of East German establishments in 1988 to 68 percent in 1995. With that, East German plants had halved the western "lead" in CAD, from 23 percentage points to just 11 percentage points.

The catch-up process of the eastern manufacturers can be seen even more clearly with the introduction of PPC systems. PPC is a technology that was highly targeted by East German firms in the 1990s. Before the economic and monetary union, merely 3 percent of eastern firms were using PPC systems in 1998, whereas the rate climbed to 61 percent in 1995. Compared to firms in the old Länder, the diffusion gap went from being 23 percentage points behind to a close lead (of one percentage point) by the mid-1990s.

128

Figure 11-1   Diffusion of selected manufacturing technologies in the old and
new federal Länder

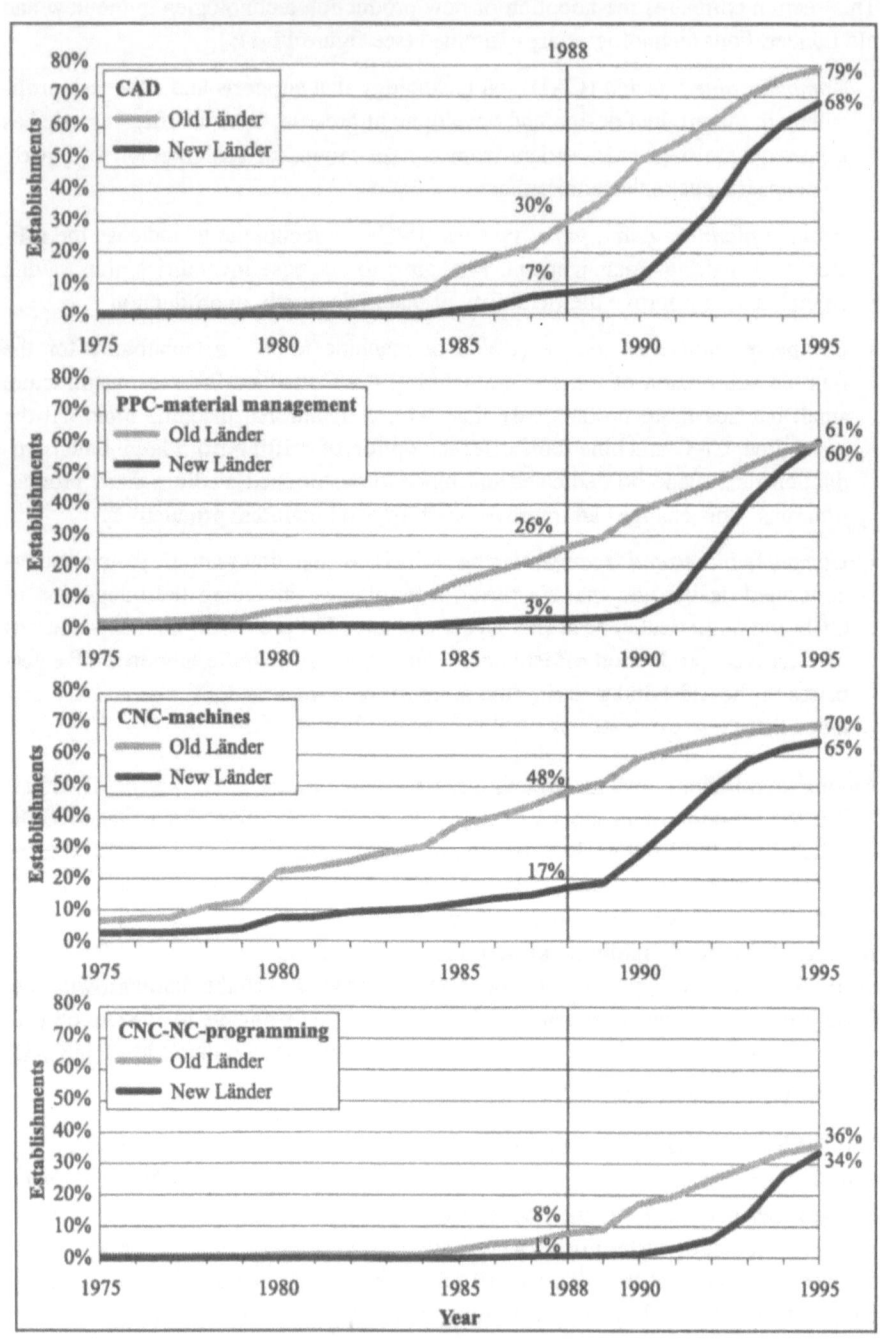

Of the modern technologies dealt with here, CNC machine tools were the technology most widely diffused during the era of the GDR. Some 17 percent of GDR plants were already using CNC machine tools before German economic and monetary union. However, at that time, almost half of the western firms were using CNC technology – a significant greater rate of diffusion. But by 1995, nearly 65 percent of East German manufacturers were using CAD, compared with 70 percent in the west. The initial gap of 31 percentage points had thus been closed to one of only 5 percentage points.

CIM – measured here by plants reporting that they integrate CAD and NC – was used by relatively few plants in both the west and east in the late 1980s. Before economic and monetary union, CIM had been adopted by 8 percent of western plants compared to less than one percent in the east. By 1995, the share of enterprises using CIM in the old Länder reached 36 percent and 34 percent in the new Länder. There is no real western lead as far as this technology is concerned.

In aggregate terms, it is clear that there has been a remarkable catch-up by east German firms in using modern production technologies since reunification. At the time of German economic and monetary union, CNC machine tools, CAD and PPC systems were far less used in eastern plants than those in the west. CIM was a technology that was only beginning to diffuse even in western enterprises, although even then there was a western lead in the use of this technology.
By 1995, the diffusion rates for PPC, CNC and CIM in the new and old Länder were comparable. For CAD technology, the east German growth rate has been faster than in the west, although as yet eastern firms have not yet caught up with western CAD usage rates.

11.3   Manufacturing Productivity in East Germany

The results described above show that, since reunification, usage rates for modern manufacturing technologies for plants in the new Länder have reached a level almost equivalent to that in the west. The technology diffusion gap evident prior at the end of the 1980s has now almost gone. However, while new technology has been deployed rapidly in the east, productivity has not increased in parallel. Official statistics suggest that in the mid-1990s, West German firms have productivity rates 45 percentage points higher than found in the east. This large difference is confirmed in the ISI survey results, based on calculations of value-added per employee, which is a standard measure of productivity. (Value-added itself is calculated by subtracting the cost of purchased input materials and services from final sales revenue.)

The ISI survey results indicate that the average value added per employee in 1995 was 133,000 DM for West German manufacturers. For manufacturers in the new Länder, the corresponding mean was only 74,000 DM. Thus, eastern plants achieved a productivity rate that was just 56 percent of the western level. The assumption that east German manufacturers could match west German performance in productivity simply by investing to the same levels in new production technology is thus not confirmed (see also Lay et al. 1996, p. 42, Mallock 1996, p. 216, Mallock and Fritsch 1995, p. 10, DIW 1997, p. 57).

What are the reasons for the continuing lag in East German industrial productivity? Clearly, other elements (besides modern technology) affect the development of efficient production structures (Ragnitz 1997b, p. 3f). The possible causes for poor east German performance can be hypothesized to include some or all of the following:

• lagging market demands which do not allow the full use of capacities

• a personnel structure operating with higher overheads compared to the west

• a lower density of application in firms using a technology

• poor organizational and inadequate use of up-to-date management methods

• the manufacturing of more complex products for which the market does not pay the corresponding prices.

In the following section, ISI survey evidence on each of these hypotheses is discussed.

East German factory capabilities and associated manufacturing workers were reduced drastically after the economic and monetary union. However, it is still possible that remaining capacities and workers are not be fully used, given difficulties in achieving adequate sales volumes in the market. A lower productivity compared to the west could thus be a result of an apportionment of the value added to plants and workers who were under-used. If this thesis were true, then the load factor of western plants with higher productivity would have to be clearly higher than in eastern plants with much lower productivity. In 1992, a survey by Fritsch and Mallock (1994, p. 55 ff) suggested that a gap in productivity in eastern firms of 55 percent was more than one third explained by lower capacity utilization (27 percent) in the east. However. by 1995, this factor no longer seems to be significant. The corresponding data from the ISI survey shows that capacity utilization between eastern and western manufacturers differs only by 4 percent (see table 11-1). This difference is too small to help explain the gap in productivity.

Even though capacities are utilized to an almost comparable extent in an east-west comparison, it is conceivable that manufacturers in the new Länder have indirect operational areas that still have excess staff capacity in a form not to be found in western manufacturers. In this case, too, the value added in the east would be spread over a higher number of employees and result in a lower productivity. A comparison of personnel structures in eastern and western manufacturers could contribute to testing this thesis. It could find support if the average share of personnel working in administration, planning departments, or maintenance proved to be higher in the manufacturers in the new Länder than in the west. Again, table 11-1 shows that this is obviously not the case. The personnel structures in eastern and western are largely similar. Establishments in the new Länder do not work with a higher overhead.

Although eastern manufacturers now use modern production technologies at similar rates to western manufacturers, it is possible that the scale of application of technology differs. For example, an eastern plant may have CNC, but only uses this technology for a small fraction of its potential CNC work tasks. On the other hand, a western plant may use CNC for manufacturing all of its parts. The ISI survey asked technology users to indicate the extent to which they actually used a new technology as a proportion of the potential application in the plant. The results, which are reported in table 11-1, show that density of use does not help to explain the differing productivity rates in east and west. On average, a plant in the new Länder using a certain technology has reached the same intensity of deployment in the potential range of application given for this technology as a western plant.

Since the late 1980s, the *organization* of production, including the way in which new technologies are operated within a plant, has been highlighted as a major factor that affects productivity. The introduction of teamwork, the clustering of manufacturing operations by products instead of by function, and simultaneous engineering are among the organization principles said to improve performance. Have manufacturers in the new Länder taken up these concepts to a lesser degree than those in the west and have they fallen behind in productivity because of this? Table 11-1 shows that this explanation also has little support in the data. New production concepts are just as widespread in East Germany's investment goods industry as in the old federal Länder. Further east-west comparisons show that the breadth of diffusion in manufacturers using new organizational concepts does not differ.

A last hypothesis that can be tested indirectly is that East German manufacturers cannot secure comparable prices in the market to those realized by western companies, perhaps because of different product mixes or market perceptions of product performance. The observed lower productivity in the east therefore would not be a consequence of a lower production output per employee by of an "under-valuation" of output. If this were true, the sales turnover and related profits of manufacturers in the new Länder would be lower than those of western manufacturers because eastern manufacturers can only find markets through lower "dumping" prices. A hint that this could be correct is found in the data. Manufacturers in the new Länder produce a remarkably higher proportion (63 percent vs. 49 percent) of customer-specified products compared to those in the old Länder (table 11-1). Customer-specified products are generally more sophisticated than products that are made with few or no customer variations. Conceivably, manufacturers in the new Länder are being required to provide extra features to customers without being able to secure higher prices to cover the extra costs involved.

Table 11-1   Comparison of productivity-relevant factors in East and West German establishments

Factor	Establishments in the	
	New Länder Percent	Old Länder Percent
Degree of capacity use	81	85
Share of personnel, by departments		
• R&D, design	10	10
• Planning	9	8
• Manufacturing	34	35
• Assembly	26	21
• Maintenance and quality assurance	5	4
• Sales and service	7	11
• Administration	9	10
Degree of using implemented technologies		
• CAD-users	67	68
• PPC- users	75	73
• CNC- users	63	72
• CIM- users	57	60
Degree of diffusion of new organizational concepts		
• Teamwork	53	53
• Segmented manufacturing	47	39
• Simultaneous engineering	25	28
• Just-in-time	17	19
• ISO 9000 certification	38	38
Manufacturing type		
• Manufacturing of customer specified products	63	49
• Program manufacturing with customer specified variations	34	46
• Program manufacturing	3	5

In summary, it appears that only one of the factors discussed seems to contribute to lagging East German productivity. Yet, how significant are these small differences in the extent of customer specified production in explaining East German manufacturing productivity? And, how does new technology use interact with productivity? To investigate these follow-on questions, the productivity rates of eastern manufacturers were divided into groups of those that

- are using modern production technologies or not, and those who

- do customer-specified manufacturing or program manufacturing with customer-specified variations.

The results show that while technology-using eastern manufacturers achieve productivity rates of between 57 to 59 percent (depending on the technology line measured) of the western average. The productivity rates of technology non-using eastern manufacturers are lower, at 51 to 55 percent. These differences are quite marginal. Thus, the use of new technology or not does not seem to be a determining factor in the explanation of the East German productivity gap. A somewhat higher relevance can be found in the east-west differences by production type. While eastern manufacturers with customer-specified manufacturing reach only 51 percent of western average productivity, the corresponding eastern figure for manufactures that run production programs with fewer customer-specified variants is 67 percent. The weakness in East German manufacturing productivity performance is thus greatest among plants that make special customized products.

11.4   Conclusions

The fact that manufacturers in the new federal Länder gain only 56 percent of productivity compared to western manufacturers in spite of similar capacity load is a matter that calls for further investigation. In the past, differences in the age and efficiency of the manufacturing tools were named as reasons for below par productivity in East Germany. Such explanations can now widely be ruled out after the enormous investment boost that moved the technology situation of East German manufacturers alongside western standards in many fields. This rapid technological catch-up process has placed major financial burdens on East German firms, depressing profits by high repayments and levels of depreciation. Yet, so far, the returns to East German technology investors in terms of increased productivity are not much greater than for those firms who did not invest.

One problem for East German firms that have modernized their technology could be that modern technologies and the ability to handle these new technological solutions productively have not developed simultaneously. In other words, the rapid pace of modernization has not allowed management and workforce capacities at all levels to fully adapt to take advantage of what these technologies can do. For example, Paasi (1997) argues that the unbalanced development of technological and management competence in East German firms has been responsible for slow progress in the economic development. Meanwhile, Mallock and Fritsch (1995) suggest that more emphasis is needed on intelligent strategies to improve the productivity gained from existing technologies, as opposed to adding more advanced capital assets.

The implication here is that a *slower* growth in adding advanced capital assets and new technological facilities could lead to *faster* progress in productivity if it is accompanied by the parallel development of organizational production concepts, improved competence in technology management, and staff training on the new technological-organizational production structures. If this thesis holds true, it suggests far-reaching consequences in management strategies for firms in the new Länder, as well as for the economic and technology policy. Such a reorientation of policy measures is an urgent task because promotion schemes for the new Länder are unlikely to be reduced due to the difficult economic conditions in East Germany (Ragnitz 1997a, 15, DIW 1997, Penzkofer and Schmalholz 1996, p. 13). Thus, better use needs to be made of available incentives. But this reorientation is unlikely to be easy, as the fascination with hard machine technology as a problem-solving strategy still seems to be very strong among East German firms. In contrast to the old Länder, East German firms continue to believe that improved efficiency in production will come about mainly through further additions of new technology (Lay et al. 1996, p. 39f). However, the results in this chapter indicate that new technology, by itself, is not sufficient to achieve this goal.

11.6   Bibliography

Dietrich, V. (1997), Kapitalausstattung und Produktivitätsrückstand im ostdeutschen Unternehmenssektor, *Wirtschaft im Wandel*, 7, pp. 5-9.

DIW (1997), Gesamtwirtschaftliche und unternehmerische Anpassungsfortschritte in Ostdeutschland, *Wochenberichte*, Deutsches Institut für Wirtschaftsforschung, 3, pp. 45-64.

DIW (1995), Nach wie vor große Defizite beim ostdeutschen Kapitalstock, *Wochenbericht*, Deutsches Institut für Wirtschaftsforschung, 31, pp. 535-544.

Dreher, C., Fleig, J., Harnischfeger, M., and Klimmer, M. (1995), *Neue Produktionskonzepte in der deutschen Industrie: Bestandsaufnahme, Analyse und wirtschaftspolitische Implikationen*, Physica-Verlag, Heidelberg.

Fritsch, J. and Mallock, J. (1994), Die Arbeitsproduktivität des industriellen Mittelstandes in Ostdeutschland – Stand und Entwicklungsperspektiven, *Mitteilungen aus der Arbeitsmarkt– und Berufsforschung*, 1, pp. 53-59.

Felder, J. et al. (1995), Innovationsverhalten der deutschen Wirtschaft – Ein Vergleich zwischen Ost- und Westdeutschland, Zentrum für Europäische Wirtschaftsforschung, Dokumentation No. 95–03.

Gürtler, J. (1997), Neue Bundesländer - Produktionspläne, Beschäftigungstendenzen und Ertragslage im verarbeitenden Gewerbe, *IFO-Schnelldienst*, 14, pp. 3-10.

Lay, G. and Mies, C. (1997), *Erfolgreich reorganisieren - Unternehmenskonzepte aus der Praxis*, Springer-Verlag, Heidelberg.

Lay, G., Michler, T., Gagel, S., and Dreher, M. (1996), *Produktionsstrukturen in der Investitionsgüterindustrie Sachsens – ein Vergleich mit den alten und neuen BundesLändern*, im Auftrag des Sächsischen Staatsministeriums für Wirtschaft und Arbeit, Sächsischen Staatsministeriums für Wirtschaft und Arbeit, Dresden.

Lay, G., Michler, T., and Pleschak, F. (1995), Angepaßte Gestaltung von CIM-Projekten - Beispiele aus der CIM-Förderung in den neuen Bundesländern, *CIM Management*, 11, 4, pp. 55-58.

Lay, G., (Ed.) (1995), Strukturwandel in der ostdeutschen Investitionsgüterindustrie, Physica-Verlag, Heidelberg.

Lay, G. and Michler, T. (1990), Produktionsautomation in der Bundesrepublik Deutschland, *Fortschrittliche Betriebsführung und Industrial Engineering*, Nr. 2, April, pp. 78-82.

Mallock, J. (1996), Kostensparende Modernisierung im ostdeutschen Maschinenbau, *ZWF*, 91, 5, pp. 216-218.

Mallock, J. and M. Fritsch (1995), Die 'Intelligenz' der Techniknutzung – Zur Bedeutung des Maschinenparks und seiner Einsatzweise für die betriebliche Leistungsfähigkeit, *Freiberger Arbeitspapiere*, 21.

Ostendorf, B. (1995), *Perspektiven industrieller Entwicklung? Transformationsprozesse des ostdeutschen Maschinenbaus*, Arbeitspapier Z2 – 5 /1993 des Sonderforschungsbereichs 187, Ruhr-Universität, Bochum.

Paasi, M. (1997), Technologische und ökonomische Kompetenzen der Unternehmen - der (noch) schwache Motor im ostdeutschen Wachstum, *IFO-Schnelldienst*, 17-18, pp. 36-43.

Penzkofer, H. and Schmalholz, H. (1996), Innovationstätigkeit und Aspekte ihrer Förderung in den neuen Bundesländern, *IFO-Schnelldienst*, 9, pp. 6-13.

Ragnitz, J. (1997a), Wirtschaftspolitischer Handlungsbedarf in Ostdeutschland - Ein Überblick, *Wirtschaft im Wandel*, 2, pp. 13-16.

Ragnitz, J. (1997b), Zur Produktivätslücke in Ostdeutschland, *Wirtschaft im Wandel*, 7, pp. 3-4.

Wengel, J. and Harnischfeger, M. (1995), Stand und Entwicklung des Rechnereinsatzes in der Produktion. In: G. Lay (Ed.), *Strukturwandel in der ostdeutschen Investitionsgüterindustrie*, Physica-Verlag, Heidelberg, pp. 78-101.

12 Promotion of CIM in the New Länder

Jürgen Wengel

12.1 Introduction

Following the reunification of Germany in 1990, companies in the new Länder of the former East Germany suddenly had to confront the cold waters of the market economy and the full convertibility of the old East German currency to the West German Deutschmark (DM). Despite many measures to promote necessary adaptation processes – including active privatization through the Treuhand privatization management agency, tax-based investment incentives, and direct subsidies – the new Länder continue to face major structural economic difficulties. East Germany's immersion in cold market waters has not yet resulted in a garden lake surrounded by a blooming economic landscape. In particular, the region's manufacturing industry has struggled. Since 1990, the new Länder experienced dramatic declines in their industrial bases. The number of workplaces in manufacturing dropped to less than one third. An initial surge in the overall growth rate, backed by transfers from West Germany, promised the eventual converging of living standards across Germany. But, this trend has recently reversed. Growth in East Germany again lags behind the West.

Against this background, the following question arises: what has been the effectiveness of the various public measures to help industrial restructuring in East Germany? The challenge of reunification was used to justify a policy of direct subsidies that would otherwise not have been in line with European Union policies and other agreements on free trade. The indirect-specific promotion of computer-integrated manufacturing (CIM), with a budget of 100 million DM, was one of these measures aimed at the modernization of manufacturing in the new Länder (and quite an important one). The Fraunhofer ISI manufacturing innovation survey provides an empirical basis for the evaluation of this measure and the following article will present some of the key findings. Due to the characteristics of the CIM technology and the CIM program it also reveals something about the impact of "soft" factors in (process) innovations and its public promotion which have recently been increasingly emphasized in technology policy debates (Lundvall 1988, Dreher 1997).

After a description of the CIM program, its context, and justification for the new Länder, this chapter addresses three sets of questions. First, what types of companies did the program reach? Was it broadly targeted or were there any discriminating factors that narrowed the kinds of companies served? Second, what impact did the different incentives and requirements of the indirect-specific promotion instrument have on company project management? Third, what impact did the program have on the implementation of CIM technologies and the economic performance of the assisted firms? After analyses of these questions, the chapter then offers some final conclusions on the CIM program.

12.2   The CIM Program and Its Context in the New Länder

The German Democratic Republic (GDR) entered economic, monetary and social Union with the Federal Republic of Germany on 1 July 1990, after some forty years of a planned economy and integration with the economies of the Soviet bloc (COMECON). East Germany's framework conditions during reunification and shortly afterwards, especially with respect to the standard of manufacturing, have been described and analyzed in a number of papers (for example, Ostendorf 1993, Britzke 1990, Haase 1990, Hirsch-Kreinsen 1992, Voskamp and Wittke 1990). These details will not be repeated here. But two general observations are important.

First, there was a common perception at that time – particularly in the new Länder themselves – that East German industry suffered primarily from a technology lag, being behind both in the extent and in the quality of new manufacturing techniques in use. This "technology lag" was believed to result in low productivity, which in turn was a major contributor to East Germany's poor economic performance. (Today we know that "technology lag" was then only partly the reason for poor productivity in East Germany and, indeed, there is no longer a significant gap in technology use, as chapter 11 discusses.)

Second, it was noted that, with respect to the qualification of the East German workforce, manufacturing units typically contained a high proportion of engineers and skilled workers. In formal terms, the organization of work was highly "Tayloristic" with centralized management structures and detailed divisions of tasks among the workforce. Nonetheless, some authors saw a base here for successful modernization. It was felt that higher level technical skills could be coupled with the abilities of the workforce to informally cooperate and improvise which had been developed on the shop floor level under conditions of the every day problems of the planned economy (Senghaas-Knobloch 1992).

Several paths seemed possible for the reconstruction of East German manufacturing structures at that time. They may be summarized under three headings:

- A traditional approach that would involve rapid technical modernization while retaining old (Tayloristic) organizational philosophies. This strategy would build on the strong technical orientation of East German managers and would address their sense of lagging behind technologically (see Vosskamp and Wittke 1990, Lange 1991).

- A leapfrog strategy, using both technical and organizational approaches to push East German firms to the forefront of modern manufacturing structures (Kern and Vosskamp 1994, Wittke et al. 1994).

- A progressive ("third way") strategy, which aimed at completely new manufacturing structures combining the above mentioned abilities on shop floor level, the adoption of new technologies, and the deployment of advanced "progressive" management and workforce experiences from West Germany (Helfert 1991, Hirsch-Kreinsen 1992).

An earlier manufacturing innovation survey carried out by ISI in 1993 in East Germany compared with data from the 1993 West German "NIFA Panel" study already showed a clear process of catching up. (The NIFA Panel study examined technical and organizational processes in mechanical engineering – see Schmid and Widmaier 1992.) However, these trends had not yet stabilized. At least the first two of the above three paths still seemed realistic (Dreher and Wengel 1994). In addition, later analyses re-adjusted the importance which was given to the technical modernization of the capital stock of East German manufacturing industry, but pointed to the loss of old markets, entrance barriers to new markets and emphasized the priority of developing company and market strategies. Unless such strategies exist, appropriate guidelines for technological modernization are also missing. Given the high uncertainty about long-term, reliable product and supply structures at that time, early and comprehensive investments risked fixing wrong (and possibly oversized) manufacturing capabilities (cf. Lay 1997).

However, when the first decisions had to be taken on the instruments to help the restructuring of the companies in the new Länder, the picture of the situation and trends was not yet as clear as described above. The belief in technical modernization of the manufacturing structures, as a key to improved competitiveness and productivity, was still predominant. At the same time, any public measure had to be broadly targeted and quickly implemented. Against this background, a number of former or current West German schemes which had been accepted by the EU already and proven largely effective were re-launched and respectively extended to the new Länder.

Figure 12-1   Indirect-specific promotion of computer integrated manufacturing (CIM)

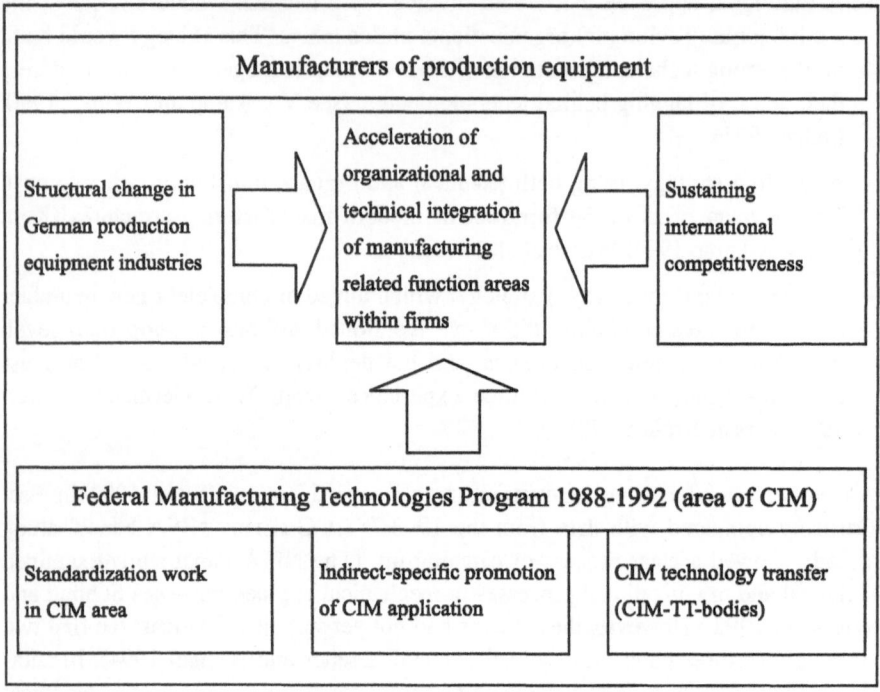

Among these was a program for the indirect-specific promotion of computer integrated manufacturing (CIM). This measure was first introduced in the framework of the Third Manufacturing Technologies Program (1988-1992) of the Federal Ministry of Research (BMBF) (see figure 12-1) but already had a predecessor in an earlier CAD/CAM program (for more information on these programs and their rationale, see Lay 1993). The indirect-specific promotion concept had formerly been chosen to make the support fast-working, ensure a broad impact and make it easily accessible to small and medium-sized firms. All available funds were allocated, while they lasted, to all applicants from the target group (manufacturers of production equipment) undertaking CIM projects on a first-come, first-served basis. Support was 40 percent of project costs, with a ceiling of DM 300,000 per firm. Costs eligible for subsidies during the planning and realization phases were internal personnel costs for the preparation and implementation of CIM concepts. Also eligible were consulting costs, training of employees, research and development contracts and software (limited to networking tasks and the replacement of old, non-integratable systems).

The scheme was transferred to the new Länder and East Berlin by the end of 1991. Ironically, in West Germany, the initial euphoria about CIM had already vanished by that time. A purely technical view of CIM with the vision of a worker-less factory had been almost completely abandoned. Yet, only a few changes were made in the program, despite the lessons from West German experience, in so far they were known, and the particular conditions in East Germany. There had been a minimum requirement for West German applicants (although this rarely proved restrictive) that the firm should have at least one CIM element installed in an ex-ante assessment (Wengel et al. 1995). But no such entrance barriers were set for the new Länder. The presumed lack of advanced computer equipment in East Germany made the BMBF even extend the eligible costs to software of CIM elements like CAD or MRP systems. In addition, the criteria for the target group were weakened. And regular, frequent payments were introduced to take account of the liquidity problems of SMEs in the new Länder. Deficits observed with CIM realization during the West German CIM program, like the predominance of technology-based approaches and the frequent deficits in the integration of the overall company strategy and the CIM concept realized, did not affect support mechanisms and project requirements.

This was all the more problematical as the need to revise their organization of production was only seen by one half of the East German investment goods firms, although the existing structures had in general been Tayloristic and centralized. At the same time 90 percent of these firms saw a strong need for technical investment, according to the 1993 ISI survey in the new Länder (Lay et al. 1994, Lay and Wengel 1995). The abilities of these companies "voluntarily" to learn from the previous experiences of Western CIM users thus presented a challenge. It was one of the most important tasks of the project administration and the CIM technology transfer centers built up at five technical universities in the new Länder to support this process. There was little risk of technical modernization proceeding too slowly. Moreover, there was a need for thoughtful as opposed to hasty investment decisions, in order to avoid CIM debacles. The integration of the investment with the development of sustainable company strategies was the main problem.

The tensions of the situation described above were reflected in the general characteristics of the CIM program and the promoted projects. Enabled by a wide interpretation of the definitions of CIM software and concepts eligible for funding, promoted projects were mainly oriented at overcoming deficits still stemming from the GDR planned economy. Accounting, distribution, material requirement planning and production control then followed another logic than that required by a market economy. Thus the common technical focus was the modernization of systems used for commercial tasks, as well as complementing or implementing material and resource planning systems rather than advanced CIM solutions. The perceived urgency for action resulted in comparatively short realization horizons. The planned total project duration was less than two years, of which less than seven months were intended for planning, whereas promoted West German CIM projects lasted more than 31 months and spent 12 months on planning on average. On the other hand, the distribution of planned project costs shows the possibilities for strategic planning in that almost 50 percent were assigned to personnel and an additional 13 percent to external consulting. Until mid 1992 all available funds had been assigned to 410 projects. The last projects finished in 1996.

12.3　The Coverage of the CIM Promotion Scheme

The 410 CIM projects promoted meant that more than 10 percent of all investment goods industry in East Germany in 1992 had been reached. The size of the actual target group of manufacturers of production equipment is more difficult to assess. One estimate indicates that there were about 1,300 production equipment manufacturers with 20 or more employees in East Germany in the early 1990s (Institut für Wirtschaftsforschung Halle 1992). Thus, an assistance rate of about one third may have been achieved, which has to be regarded as high. The following analysis of the selection of the promoted firms refers to the investment goods industry either on the basis of available statistics or the 1993 ISI survey in East Germany (as it best reflects the situation at the time of the introduction of the CIM program). The structure of this sector can be assumed as very close to the actual target group. However, it is of course bigger and there are investment goods manufacturers which do not produce manufacturing equipment, as well as other sector companies with a share of their turnover in such products which made them eligible to the CIM subsidies.

Figure 12-2 gives the distribution of promoted firms with respect to their size in comparison to the investment goods industry. It clearly shows an under-representation of smaller manufacturers, with a reciprocal over-representation of medium-sized and (particularly) big establishments. Nevertheless, this distribution indicates the large impact of the program on small and medium-sized firms. The share of very small firms with less than 20 employees participating – many of them stemming from the break-up or separation of big conglomerates ("Kombinate") – was twice as high as in the comparable West German CIM program. Certainly, it should not be expected nor requested that smaller firms participate in a technology diffusion measure with a considerable R&D component like CIM to the same degree as bigger companies. Small firms are less likely to make full use of expensive advanced technology (Lay and Michler 1989). On the other hand, there are indications that they still suffer from a lack of information and ability to cope even with the comparatively simple procedures of the indirect-specific promotion concept. Whereas bigger firms primarily belonged to the early applicants, the smaller ones came late and thus many might have missed the financial support, since funds were exhausted.

Figure 12-2   Size of CIM-promoted establishments

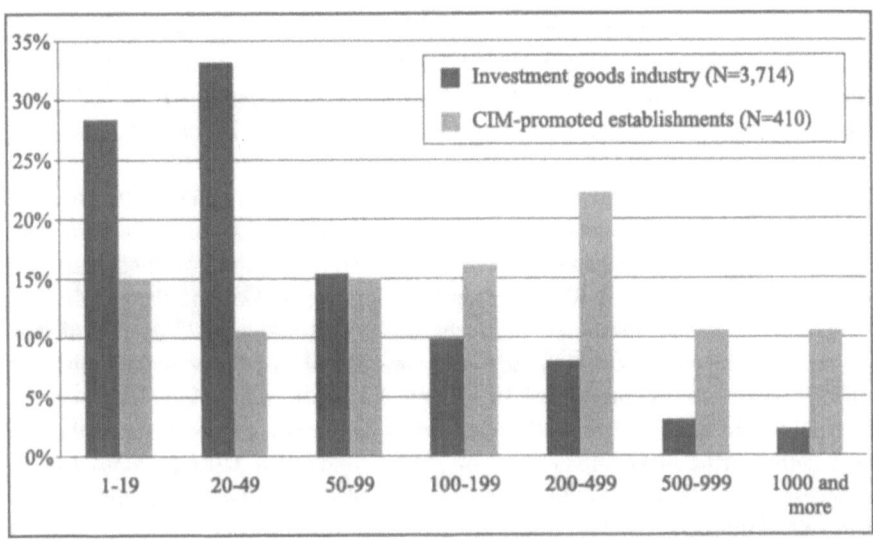

Looking at the regions where promoted and non-promoted companies were located, Saxony had the highest share of almost double as many firms as would have been expected, according to its number of investment goods manufacturers. This is partly explained by the size structure of the investment goods industry (higher share of bigger firms) and the likely overrepresentation of manufacturing equipment producers in Saxony, since this state was at the heart of the mechanical engineering sector in GDR and hosted most of the relevant industrial conglomerates. However, a reinforcing factor might have been that a branch of the project administration body for manufacturing technologies from the Karlsruhe research center (Forschungzentrum Karlsruhe – Projektträger Fertigungstechnik or FZK-PFT) was set up in Dresden. This certainly made the program more visible in the region. Additionally, compared to other East German Länder (cf. Lay and Michler 1996), Saxony from the very beginning pursued a technical modernization strategy through additional subsidies for capital investments.

The following structural differences between the program participants and the control group were identified on the basis of the ISI´s 1993 state-of-the-art survey on manufacturing in East Germany. It included 275 CIM-promoted and 513 non-promoted firms. Again they are correlated with the establishment size. But as the size seldom is a cause in itself, they might point to other underlying explanations for the up-take or non-receipt of public funding.

As the restructuring of East German industry coincided with its privatization, the ownership of the companies in transition was often a subject of discussion at that time. Our data show that firms which belonged to the Treuhand (these are predominantly bigger ones) or had West German owners made more frequent use of the CIM program than foreign and – particularly – East German owned companies. This is probably both due to a better knowledge of different public schemes on the part of Treuhand managers as compared to foreign managers, and the lack of liquidity among East German owners to afford their share of CIM project costs.

Furthermore, more innovative firms were more likely to participate in the program. This was true for product innovations as well as for process innovations. The average share of turnover of products newly introduced during the last three years was almost 10 percent higher with CIM-promoted companies. Only 15 percent of CIM program participants had no new products at all, compared with 23 percent of non-participants. The extent to which the two groups already used new technologies like MRP II, CAD, or CNC machine tools in 1991 was more similar, but on average the penetration was again larger with promoted firms. It comes as no surprise that the more active firms make greater use of public funds. The CIM program and its predecessor on CAD/CAM in West Germany also attracted the technologically more advanced firms.

With respect to the later assessment of the economic impact of the program, it is important to check previously in how far the above described selectivity also involves a better performance from the beginning in order not to mistake such head starts for effects of the promoted CIM projects. The indicators available draw a disparate picture. Higher exports of promoted companies may in general indicate better international competitiveness. But at the same time, exports may reflect the distribution patterns of the old planned economy in which the GDR mechanical engineering sector was a major supplier for other COMECON members, and the Treuhand policy to maintain these distribution channels. With respect to employment change, promoted firms between 1989 and 1992 followed the same negative trend as other investment goods producers. The cuts even seem to have been somewhat deeper. Given the low productivity of East German firms a quick adaptation of the workforce to an economically viable level could be a competitive advantage. However, value-added per employee in 1992 was not higher in promoted than in non-promoted firms over all firm size classes. Only the self-assessment of their economic situation gives a slightly more positive impression as only 33 percent of the CIM program participants vis-à-vis 39 percent of the non-promoted investment goods manufacturers saw threats to their survival due to decreasing or stagnant turnover.

Against the above background, no clear-cut picture can be drawn of the CIM-promoted in contrast to the non-promoted companies. This seems to be due to the partly contradictory characteristics of the two biggest groups of program participants, both over-represented:

- The mainly medium-sized East German branches of West German firms carried out a moderate technical modernization policy, and did comparatively well.

- Treuhand companies not yet successfully privatized which were usually larger, as the core part of former conglomerates had installed some advanced technical equipment, and invested more heavily in technical modernization, but suffered economically.

12.4 Project Management in CIM-Promoted and Non-Promoted Establishments

The 1995 ISI survey of innovation in production covered issues of project management in two ways. First, assisted manufacturers were asked for a self-assessment of the impacts on their project management behavior and the shaping of their CIM projects. Second, all survey participants were asked to give detailed information on a recent comprehensive techno-organizational project, which in the case of the promoted establishment was pre-determined as their CIM project, thus allowing for comparison with non-promoted West and East German establishments which carried out similar activities.

The promoted establishments were aided through several means, including assistance planning procedures, training, or external support by consultants, software or hardware suppliers, system houses, or research bodies. But, irrespective of the form of program assistance, almost all of the 216 promoted manufacturers included in the survey reported an impact on their respective project management behavior (figure 12-3). In the field of planning, it apparently would have been difficult for many East German firms to provide any own personnel capacity. More than 40 percent mentioned this. Almost one third said that they would not have examined alternative solutions without the public funding. But with 8 percent of the cases the opposite was also true, that they looked less intensively for alternatives. Probably the additional funds allowed a pre-planned solution. The main effect of the promotion, however, was that the firms intensified their planning work. Due to planning, firms reported changes of emphasis in the four categories of company strategy, technical questions, organizational aspects, and personnel development.

The impact of program participation on training-related activities can also mainly be characterized as intensification, with the exception of the appointment of persons and assignment of capacity to elaborate a training concept. Almost 40 percent of the promoted firms would not have done this without public funding. The CIM subsidies seemed also to have very much helped the development of know-how in the organization and with respect to the methods of training. These competencies can certainly be useful for later modernization measures. However, the share of compa-

Figure 12-3   Impact of promotion on CIM project management

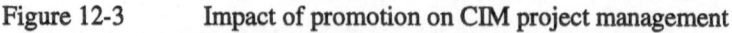

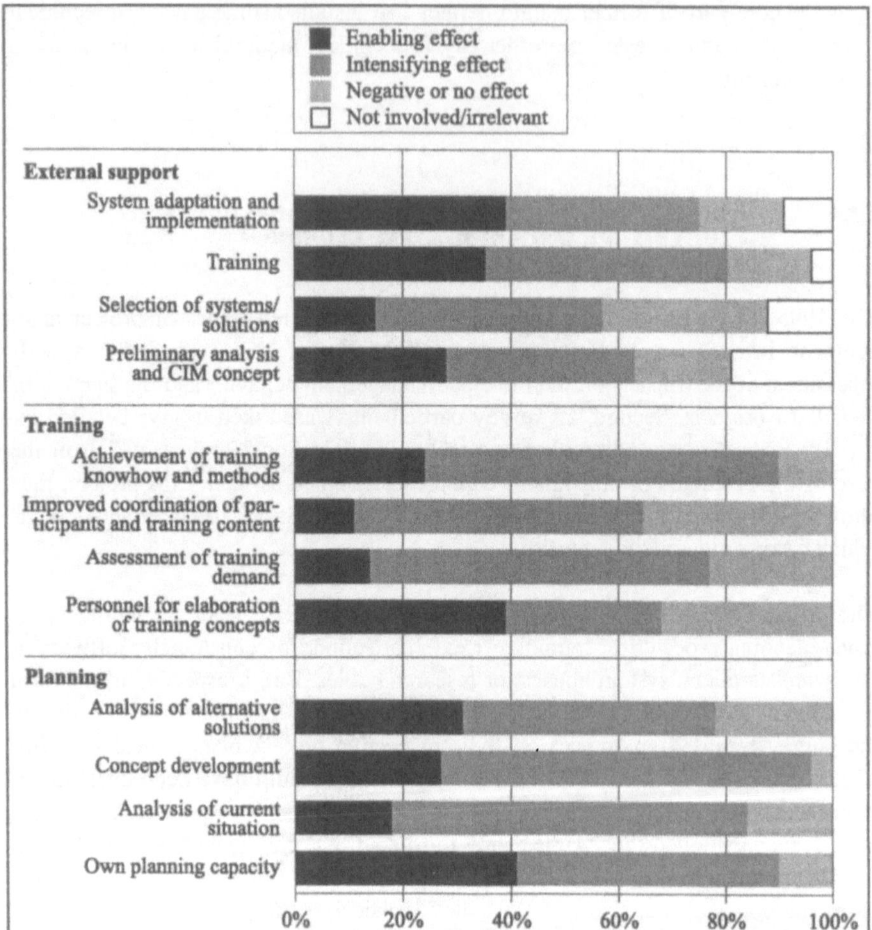

nies reporting no effects is higher with training. Results from case studies (Uhr-mann et al. 1997) imply that high motivation and high basic skills of the workforce in many cases allowed and were used for "minimum" training strategies. If (mini-mum) training is just seen as a necessary precondition rather than an important suc-cess factor of process innovations, even additional funds will hardly change such attitudes as long as they could be spent on other items in the promoted CIM project as well.

A number of firms reported a "negative" effect or decrease in planning or training efforts due to the public funding. The explanation usually lies in the shifting of such efforts to external experts. The eligibility of such costs for subsidies obviously enabled as well as increased the use of external support, particularly with respect to training (36 percent respectively 46 percent) and to the development, adaptation, or implementation of software (39 percent respectively 36 percent of the firms). External expert contributions seemed to have been quite important. Every tenth firm believed that the external experts decisively shaped their CIM project. Every second saw therein a major contribution. In little more than one third of the promoted firms the support concerned only small tasks. Only five percent of program participants did not involve external experts at all in the planning and implementation process.

The self assessment of the firms of a strong impact of the program and its specific support regulations on their project management behavior are underlined by the comparison of promoted and non-promoted projects as they were described by the firms participating in the survey. In order not to mix non-comparable activities, the projects selected for this analysis are ones which focused on core CIM elements like MRP II, CAD, CAM, shop floor data collection and the related interlinks. At the same time, the structural differences between promoted and non-promoted firms (see section 12.3) have to be taken account of.

Figure 12-4 shows that the establishment of a project team, which was required by the CIM program, is not self-evident even with a complex task like the implementation of CIM. Whereas almost all of the promoted firms mentioned a team to manage the project, only two thirds of the non-promoted East German firms used the team organization for their CIM projects. The non-promoted West German firms as well make use of teams to a lesser extent than the CIM-promoted, although their higher average size would make that more sensible. With respect to the understanding of CIM as a re-organization measure using CIM technology as a tool, as opposed to an investment with some accompanying organization and training activities, the funding was not that successful. Despite the disillusionment about technically focused CIM in the West and the information measures accompanying the program in East Germany, only 30 percent of projects emphasized the organizational focus, according to our survey results. By comparison, at least 40 percent of West German firms pursued an organizational orientation. At the same time, the share of pure reorganization projects, e.g. to implement employee teamwork (which was not included in the comparison), was much higher in West Germany. However, in the non-promoted East German firms the technical orientation dominated even more. In this group, only 20 percent chose an organization-oriented approach.

The duration and volume of the projects primarily reflect firm sizes, economic conditions, and specific priorities in the realization of CIM. Thus, a comparison is difficult. Predictably, subsidized firms could allot more financial resources (30 percent on average) to their CIM projects than non-promoted East German ones (though they could not achieve average West German financial efforts). The overall project duration was considerably longer with promoted firms, both in comparison to non-

Figure 12-4   CIM project management structures, non-promoted establish-
                     ments compared with promoted establishments

Percent difference from base line of promoted establishments

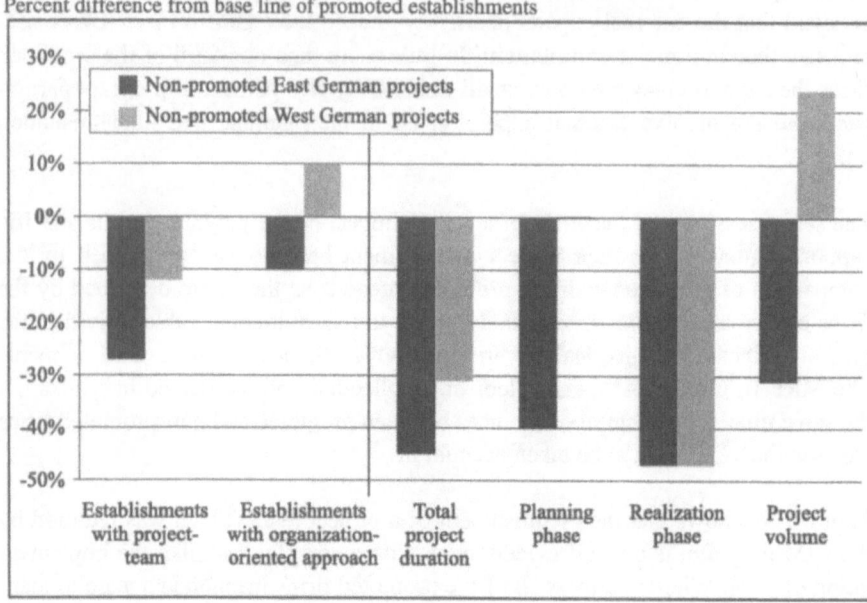

assisted East German manufacturers and the West German counterparts. Given the
largely uncertain circumstances at that time for long-term decisions on future
manufacturing structures, the program-inherent possibilities of stretching the CIM
investments might have proved an advantage for subsidized firms. Certainly, the
CIM subsidies provided additional liquidity to keep qualified experts in the promoted firms. The cost structure (see figure 12-5) of the promoted projects supports
this assumption, as more than 40 percent of the costs went into planning by own
personnel, whereas non-promoted firms spent around two thirds and three quarters
respectively on hardware and software. Higher shares of expenditure for training
and external experts back up the above assessment of positive impact of the CIM
promotion measure in these two respects.

12.5   The Impact of the Promotion on the Firms

The original prime goal of the promotion program was to speed up the diffusion of
CIM technologies in the new Länder, thus improving the productivity and consequently the competitiveness of the participating firms. However, it was already
mentioned (and the previous chapter 11 of this book underlined it) that a straightforward strategy of technical modernization would not generally have been the best

Figure 12-5   CIM project cost structures in promoted and non-promoted establishments

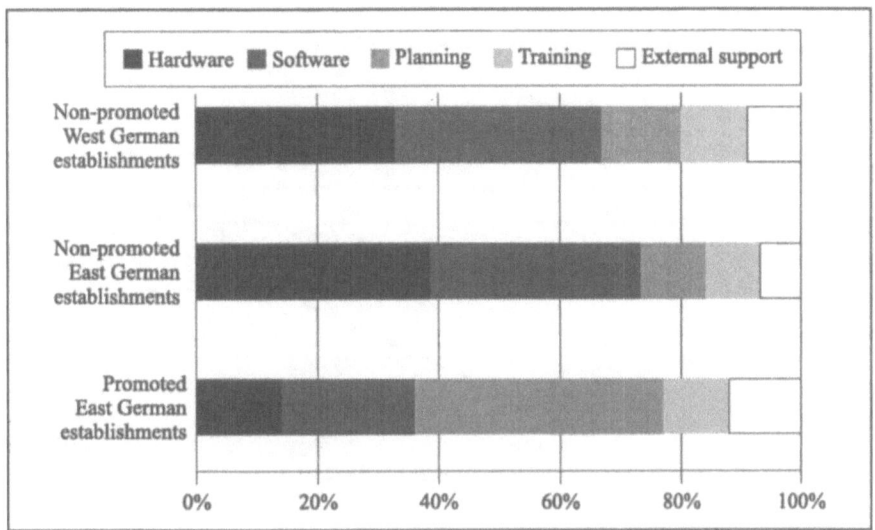

approach in East Germany in the early 1990s. Rather, a moderate, very sensible introduction of CIM technologies that continuously reflect changing market requirements and their impact on the overall company strategy was required. Arguably, heavy investments in advanced technology would have occurred anyway, given the feeling of "lagging behind" common to responsible East German production managers.

As it turned out, the effects of program participation are somewhat contradictory. On the one hand, only very few companies (2 percent) admitted they would have done their CIM project in the same way without any funding, according to the self assessment of the impact of subsidies on firm behavior in the 1993 questionnaire. A larger group (18 percent) only started the CIM project because of available public funds. Most establishments (61 percent) reported that additional funding prompted expanded effects, meaning wider or better application of CIM elements and techniques. On the other hand, a comparison between the promoted and non-promoted firms of their adoption dynamics also suggests that CIM program participants benefited from a "breathing space" during the most critical phase of technical investments (see figure 12-6). Shortly after the monetary union and the re-unification many firms (apparently) almost blindly implemented now available, new western technologies very quickly, whereas in many cases the CIM subsidies seem to have provided the time and capital buffers for a more thoughtful introduction strategy.

Figure 12-6   Dynamics of CIM implementation in promoted and non-
                   promoted establishments

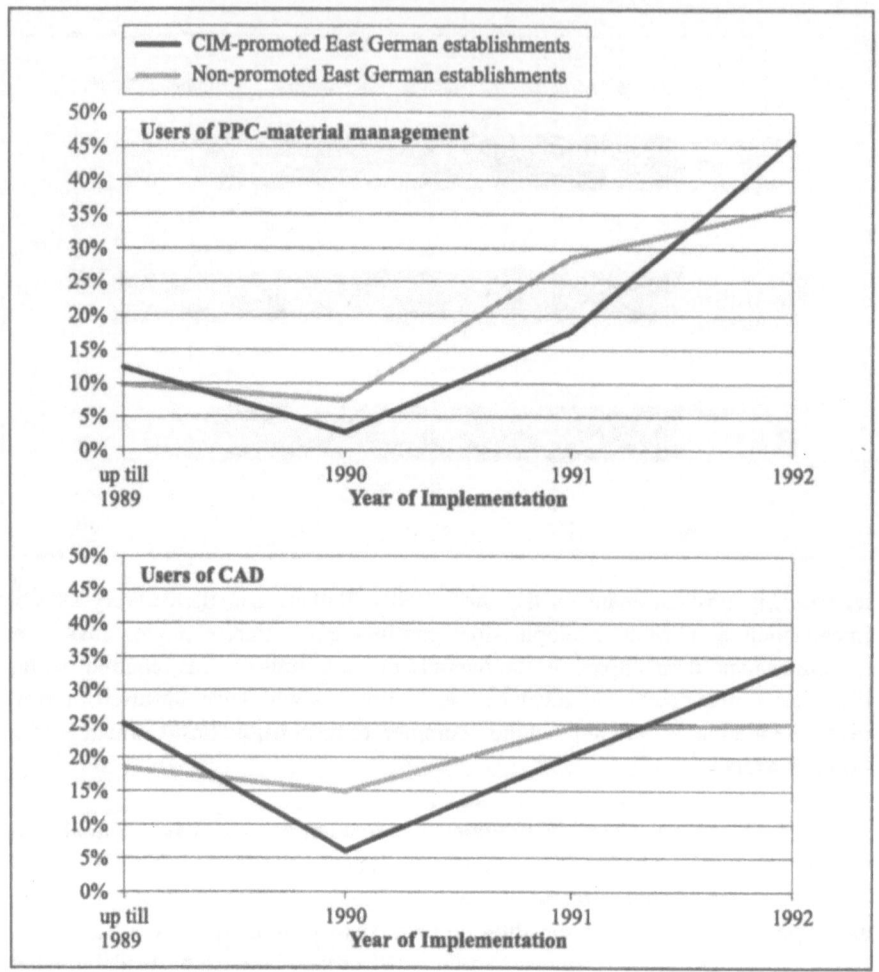

Figure 12-6 illustrates the diffusion of computer-integrated manufacturing methods, by showing the year of first implementation of CAD technology and PPC (production, planning, and control) systems using materials management (specifically, material requirements planning, or MRP). That almost 50 percent of the non-promoted compared to just one third of the promoted firms had introduced MRP by the end of 1991 is a clear indication that the program – rightly – moderated rather than accelerated the diffusion of CIM in the beginning. This is more significant when we consider that non-participants were much smaller on average and therefore would have had to be even more careful not to overdo CIM investments.

Subsequently, as the 1995 survey shows, participants of the CIM program achieved a wider application of almost all CIM technologies, as well as of new production concepts like simultaneous engineering, R&D teams, continuous improvement processes, or employee teamwork even though these were not directly targeted in the CIM program. Possibly, the overall strengthening of the planning and implementation capacity in assisted firms also helped with the adoption of new organizational arrangements.

Finally, the CIM promotion measure was intended to improve the competitiveness of the participating firms. Under the circumstances of a complete restructuring of East German industry as a whole and of each firm individually, it is particularly difficult to assess the contribution of the CIM projects and the resulting improvements of technological, organizational and management capacities of the firms on their economic performance. Many activities – in particular with respect to personnel – were undertaken in parallel (cf. Uhrmann et al. 1997) and not least, virtually no firm ever tried to evaluate the outcome or cost-benefit-relation of their CIM activities (cf. Pleschak 1997). Against this background, the comparison of promoted and non-promoted firms on the basis of the survey can only give some hints on the respective economic performance and the benefits gained. What is cause and what is effect is hard to determine; there is no chance of full control of the numerous influencing factors among the two groups as well as uncertainty about the time when the benefits of the modernization of the manufacturing structures via the CIM projects may occur.

The survey on the one hand asked for direct effects in connection with the project, i.e. problems during the projects and an assessment how far the set goals had been met. With one exception, the promoted companies reported less or just the same extent of problems. The exception was the final realization of the initial CIM concept. One quarter of the promoted but only every tenth of the non-promoted companies said that they could not realize the CIM concept they had first in mind. Obviously, due to the extended planning considerations at the beginning on how to implement CIM underwent a more serious review in promoted firms. It can be assumed that this was to the advantage of the promoted firms, as they should in general have had the ability – through the subsidies – to readjust their approaches. With respect to cost reduction and quality, the CIM program participants significantly more often (74 compared to 69 and 79 to 72 percent) said that they achieved or even surpassed their goals. Flexibility targets had as often been met (79 percent). Only improved delivery times were slightly more often (79 to 77 percent) achieved or exceeded by non-promoted companies. As bigger companies were more critical about the effects of their CIM projects, the above figures underscore the better assessment in promoted projects among which bigger firms were over-represented.

Figure 12-7   Economic situation and employment in promoted and non-promoted establishments

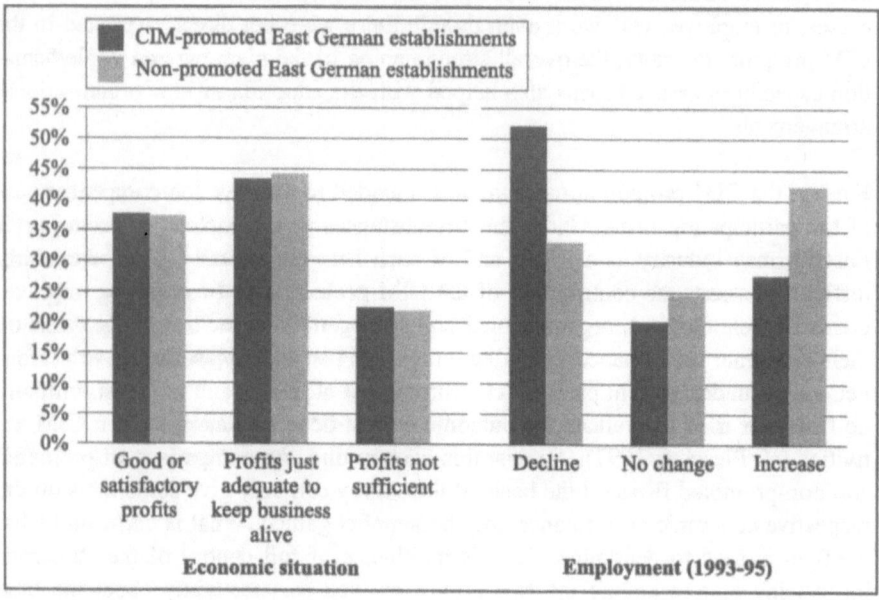

On the other hand, the survey provided a number of indicators on the overall performance of a company. Figure 12-7 compares the self-assessments of the current economic situation and the development of the employment in the last two years between promoted and non-promoted firms. It shows that promoted firms do not differ from non-promoted firms as regards their current economic situation. However, they have a much more optimistic view of their future economic performance. The employment situation reflects this only to some extent. A much higher share of the promoted firms had reduced their workforce in the last two years. And both stable and increasing employment was more often to be observed with non-promoted firms. The lower share in the CIM program of SMEs, which in general show less negative employment trends, does not fully explain this. But again the view of the future was more optimistic with the CIM program participants. Only 14 against 19 percent expected a (further) decrease in the workforce and 39 compared to 34 percent believed in a positive employment trend. One could therefore conclude that the promoted firms already walked through their "valley of tears" and that CIM technologies had been implemented along with a rationalization of employees. They apparently by a great majority see their achieved position as a good starting point for future economic success.

12.7 Conclusions

The CIM program was introduced in East Germany in the early 1990s to address a common perception that firms in the new Länder were technically lagging due to a history of limited access to state-of-the-art manufacturing equipment and information technology. The belief in technological modernization as the way towards competitiveness was widespread in East German industry, although the CIM euphoria in Western countries had already vanished. On the other hand, in most instances, medium and long-term corporate strategies to guide technical investment decisions were missing or, at least, difficult to develop. The period was one of deep parallel changes in the economy and society, with high uncertainty about the new market environment and structure of industry. This was a difficult background for a promotion measure that aimed at the diffusion of an advanced, complex technology.

The ISI survey proved very useful in evaluating the CIM program in the new Länder in that it provided the basis for differentiated comparison of promoted and non-promoted firms (control group). Unfortunately, the time horizon of the sponsors of the ISI CIM evaluation project did not allow a full long-term assessment of the impact of public funding on the performance of companies, although in principal the panel character of the survey allowed this. But it was possible to draw a series of shorter-run observations and conclusions.

The CIM program, because of the size of the funds involved, emerged a major factor in the modernization of manufacturing in the East German investment goods industry. Because of the indirect-specific promotion concept, it not only covered a large number of companies, but it also reached small and medium-sized enterprises to a considerable extent though bigger companies were nevertheless over-represented.

Several voices expressed the hope that East German companies might be able to apply a leap-frog strategy so as not to repeat mistakes already made and recognized – e.g. the vision of a worker-less CIM factory – and thus exploit new production concepts to the full. Despite the prominent featuring of CIM in the program title, the promotion concept itself provided opportunities in the above direction in that it emphasized (and financially supported) planning, employee training, or consultation with external expert. Actually, the ISI survey, not least by comparison with non-promoted firms, showed that the subsidies had a major impact on the shaping of the CIM projects and improved the project management behavior of the promoted firms. However, a technical orientation still dominated most of the funded CIM projects. While this was even more evident among the other non-subsidized East German CIM projects, a much lower share of West German firms followed a technical orientation.

Thus, while company managers can be influenced by the application and steering of appropriate policy instruments, the program's simple financial incentives, by themselves, did not change long-standing attitudes towards technology. The (mistaken) belief in technology alone as the key to better competitiveness was continued in the assisted East German firms. However, the program did affect other aspects of management behavior and decision-making. It strongly influenced decisions on how to implement CIM technologies, according to the self-assessment of the promoted firms. Only two percent said that they would have done the project in the same way without the subsidy.

At the same time, annual diffusion rates show that promoted companies gained a "breathing space" (compared with non-promoted East German companies) during a critical phase of technical investments. The CIM subsidies seem to have provided the time and capital buffers for a more thoughtful implementation strategy. Ultimately, a wider diffusion of most technologies and new organizational concepts was to be observed with CIM program participants. However, their economic performance in 1995, when just over one half of the projects were completed, did not appear significantly better than those of non-promoted firms – but their assessment of the future was far more optimistic.

12.8   Bibliography

Britzke, B. (1990), Arbeitsgestaltung und Arbeitsorganisation in den Betrieben der ehemaligen DDR, *Angewandte Arbeitswissenschaft*, 126, pp. 1-10.

Dreher, C. (1997), Technologiepolitik und Technikdiffusion: Auswahl und Einsatz von Förderinstrumenten am Beispiel der Fertigungstechnik, Baden-Baden, Nomos.

Dreher, C. and Wengel, J. (1994), How do Techno-organizational Paradigms Change after Fundamental Shifts in Frame Conditions? The Case of Flexible Automation in Eastern Germany. In: Cohendet, P. and Willinger, M. (Eds.), EUNETIC European Network on the Economics of Technological and Institutional Change: Evolutionary Economics of Technological Change. Assessment of Results and New Frontiers, EUNETIC Conference, European Parliament, Strasbourg, October 6-8, 1994, *Proceedings*, Université Louis Pasteur, Strasbourg, pp. 1123-1146.

Haase, H. (1990), *Das Wirtschaftssystem der DDR – Eine Einführung*, Berlin.

Helfert, M. (1990), Arbeitspolitische Aspekte industrieller und sozialer Modernisierung in der DDR, *WSI-Mitteilungen*, 10, pp. 668-679.

Hirsch-Kreinsen, H. (1992), Modernisierungsrisiken im ostdeutschen Maschinen-bau, *WSI-Mitteilungen*, 5, pp. 293-330.

Institut für Wirtschaftsforschung Halle (1992), *Vorschlag zur Bestimmung der Zielgruppe für die Evaluierung des Förderprogramms Fertigungstechnik des BMFT in den neuen Bundesländern*, unveröffentlichtes Manuskript, Berlin.

Kern, H. and Voskamp, U. (1994), Bocksprungtheorie – Überholende Moder-nisierung zur Sicherung ostdeutscher Industriestandorte, *Sofi-Mitteilungen*, Göttingen, 21, pp. 98-138.

Lange, H. (1991), Arbeits- und Technikgesaltung in den neuen Bundesländern – kein Thema? In: Fricke, E. (Ed.) *Konzepte zur Gestaltung von Arbeit und Technik aus Wissenschaft und Praxis*, Friedrich-Ebert-Siftung, Bonn, pp. 89-96.

Lay, G. (1993), Government Support of Computer Integrated Manufacturing in Germany: First Results of an Impact Analysis, *Technovation*, 13,5, pp. 283-297.

Lay, G. (1997), Beiträge von CIM-Projekten zur Verbesserung der Wettbewerbspo-sition ostdeutscher Betriebe. In: Wengel, J. (Ed.), *Modernisierung der Pro-duktionsstrukturen in den neuen Bundesländern*, Forschungszentrum Karlsruhe, Technik und Umwelt, FZK-PFT 184, pp. 89-107.

Lay, G. and Michler, T. (1989), *Stand und Aussichten der Fertigungsautomation in der Bundesrepublik Deutschland*, Fraunhofer Institute for Systems and Inno-vation Research, Karlsruhe.

Lay, G. and Wengel, J. (1995), Wettbewerbssituation und Handlungsprioritäten in der Investitionsgüterindustrie Ostdeutschlands: Analyse der Ausgangslage für die CIM-Förderung in den NBL. In: Kuhlmann, S. and Holland, D. (Eds.), *Systemwandel und industrielle Innovation*, Studien zum technologischen und industriellen Umbruch in den neuen Bundesländern, ISI Schriftenreihe, Band 16, Fraunhofer Institute for Systems and Innovation Research, Karlsruhe, pp. 113-130.

Lay, G., Michler, T., Gagel, S., and Dreher, M. (1996), *Produktionsstrukturen in der Investitionsgüterindustrie Sachsens – ein Vergleich mit den alten und neuen Bundesländern*, im Auftrag des Sächsischen Staatsministeriums für Wirtschaft und Arbeit, Sächsischen Staatsministeriums für Wirtschaft und Arbeit, Dresden.

Lay, G., Schmierl, K., Schultz-Wild, R., and Wengel, J. (1994), *Bestandsaufnahme der Produktionsstrukturen in der Investitionsgüterindustrie Ostdeutschlands*. Information für Betriebe, die sich an der Umfrage im Jahre 1993 beteiligt hatten, Fraunhofer Institute for Systems and Innovation Research, Karlsruhe.

Lundvall, B. (1988), Innovation as an Interactive Process: From User-Producer Interaction to the National Systems of Innovation. In: Dosi, G., Freeman, C., Nelson, R., Silverberg, G., and Soete, L. (Eds.), *Technical Change and Economic Theory*, Pinter, London, pp. 349-369.

Ostendorf, B. (1993), Perspektiven industrieller Entwicklung? Transformationsprozesse des ostdeutschen Maschinenbaus, Arbeitspapier Z2-5/93 des Sonderforschungsbereichs 187, Ruhr-Universität, Bochum.

Pleschak, F. (1997), Theorie und Praxis von Wirtschaftlichkeitsuntersuchungen für rechnerintegrierte Produktionsstrukturen. In: Wengel, J. (Ed.), *Modernisierung der Produktionsstrukturen in den neuen Bundesländern*, Forschungszentrum Karlsruhe, Technik und Umwelt, FZK-PFT 184, pp. 75-88.

Schmid, J. and Widmaier, U. (Eds.) (1992), *Flexible Arbeitssysteme im Maschinenbau*, Ergebnisse aus dem Betriebspanel des Sonderforschungsbereichs 187, Leske and Budrich, Opladen.

Seghaas-Knobloch, E. (1992), Notgemeinschaft und Improvisationsgeschick. Zwei Tugenden im Innovationsprozeß. In: Senghaas-Knobloch, E. and Lange, H. (Eds.), *DDR-Gesellschaft von innen: Arbeit und Technik im Transformationsprozeß*, Friedrich-Ebert-Stiftung, Bonn, pp. 91-101.

Uhrmann-Nowack, R., Winter-Hoss, R. and Greiner, C. (1997), Qualifizierungsverhalten in Umbruchsituationen – Personal und Qualifikation im Kontext der CIM-Einführung in ostdeutschen Betrieben. In: Wengel, J. (Ed.), *Modernisierung der Produktionsstrukturen in den neuen Bundesländern*, Forschungszentrum Karlsruhe, Technik und Umwelt, FZK-PFT 184, pp. 47-73.

Voskamp, U. and Wittke, V. (1990), Aus Modernisierungsblockaden werden Abwärtsspiralen – Zur Reorganisation von Betrieben und Kombinaten der ehemaligen DDR, *Sofi-Mitteilungen*, Göttingen, pp. 12-30.

Wengel, J. (Ed.) (1997), *Modernisierung der Produktionsstrukturen in den neuen Bundesländern*. Ergebnisse und Erfahrungen aus der Analyse von Fallbeispielen geförderter und nicht geförderter Projekte zur Realisierung rechnerintegrierter Produktion im Rahmen der Evaluierung der ostdeutschen CIM-Förderung, Forschungszentrum Karlsruhe, Technik und Umwelt, FZK-PFT 184.

Wengel, J., Lay, G. and Dreher, C. (1995), Evaluation of the Indirect-Specific Promotion of Manufacturing Technology. In: Becher, G. and Kuhlmann, S. (Eds.), *Evaluation of Technology Policy Programs in Germany*, Kluwer Academic Publishers, Dordrecht, pp. 81-99.

Wittke, V., Voskamp, U. and Bluhm, K. (1993), Den Westen überholen, ohne ihn einzuholen? Zu den Schwierigkeiten bei der Restrukturierung der ostdeutschen Industrie und den Perspektiven erfolgversprechender Reorganisationsstrategien. In: Schmidt, R. (Eds.) *Zwischenbilanz-Analyse zum Transformationsprozeß der ostdeutschen Industrie*, Berlin.

13   Implications for Modernization Strategies and Studies

Gunter Lay, Carsten Dreher, and Philip Shapira

13.1   Introduction

In recent years, the process of innovation has moved very much to the forefront of concern in understanding and promoting industrial competitiveness, business performance, and economic growth and development. Nonetheless, as the economist J.M. Keynes observed more than fifty years ago: "The difficulty lies, not in the new ideas, but in escaping from the old ones..." (Keynes 1936). This tension between innovation and convention has certainly been evident in the analyses presented in this book. The contributions to the book have examined in detail trends in the diffusion of new production concepts, taking into account such differences as underlying industrial structure, product mix, establishment size, and location factors to explain variations in adoption rates. The context has been that of the German investment goods sector, which is now challenged to change traditional structures of production and innovation under pressures of global competition and domestic restructuring.

In this final chapter, the major findings of the book are reviewed. However, the main aim in the following sections is to build on the results discussed in the book to explore issues and implications. First, we highlight the ramifications of the book's analysis for debate about German competitiveness and industrial strategy. Following this, we explore specific implications for management, policy, and future research.

13.2 German Competitiveness, Industrial Strategy and Business Performance: Recasting the Debate

The analyses in this book have offered fresh insights for the broader debate about German industrial competitiveness. In the opening chapter to the book, we highlighted several common perceptions about factors felt to impede German competitiveness. Even then, contrasting evidence was noted which suggested that these popular accounts did not tell the whole story. This proposition was verified by the analysis of the ISI manufacturing survey. In several areas, preconceived notions about the German competitive environment were found not to be accurate.

For example, it has been argued that the labor-management bargaining process involved in German codetermination may slow down corporate decision making. However, the assumption that extended employee participation leads to disadvantages for the German industry is not proved by our results. We found quite the opposite: companies that practice a participation-oriented organizational culture experience increased acceptance of process innovations and demonstrate advantages on the relevant performance indicators. The more that employees are seen as incorporating important information and knowledge and the more their creativity is engaged in seeking best solutions, the less relevant the issue of worker resistance becomes.

Second, contrary to general perceptions, we found that competitive constraints due to environmental regulation and compliance play a rather minor role in the German investment goods producing sector. Some may argue that German industry is ignoring important environmental issues and is under investing now in the environmental area, possibly to future regret and loss of potential comparative advantage. But, while there is intense public debate on environmental issues in Germany, we did not find that current industrial decision making is hampered.

Third, we found little evidence to substantiate claims about the inflexibility of German industrial workers. Indeed, it was demonstrated that flexibility in working time to rapidly meet customer requirements has passed through the experimental stage in the German investment goods industry and, for many establishments, has now become a part of everyday life. It is interesting to note that greater flexibility in working and operating hours has been accepted without drama in German industry, particularly when compared with the tense public debates on the opening hours in the German retail and service sectors governed by federal regulation. Industry is clearly exposed to international competitive pressures that have required greater flexibility.

However, while many German firms are adopters of the new production concepts, it is also apparent that many other companies have yet to make comprehensive use of the available approaches. Clearly, not all the elements of the new concepts can be effectively deployed in every firm. But, this said, why have many German firms lagged in using new production concepts? As we have emphasized, the reason for this lag cannot be attributed to a general anti-competitiveness in the German business climate, or to such factors as labor-management bargaining, regulation, or work attitudes. This is not to say that there are not constraints to the faster diffusion of new production concepts in Germany. Indeed, there are – and they include obstacles of information, expertise, time, and finance. However, such obstacles are common across all industrialized economies, particularly for small and medium-sized firms (for comparison, see Dodgson and Bessant 1996, Shapira 1996). Considering this, there is little evidence that, across the board, German investment goods manufacturers are doing less well than their competitors in major industrial countries in adopting new manufacturing technologies and organizational concepts.

However, there are issues of concern not so much about the quantitative degree of the diffusion of new production concepts in Germany, but about the business strategies and production frameworks within which new production concepts are deployed. At several points in the book, authors noted the tendency to focus debate and subsequent action in introducing new production concepts within a strategic framework oriented towards cost reduction, rather than to promote the comprehensive modernization and innovative orientation of production structures. Although, at first glance, this is an orientation influenced by Germany's high wage levels, we conclude that this is driven by a more significant and deeper set of issues related to the underlying competitive strategy being pursued by many German firms.

German manufacturers in principal have several strategic options to react to global production, changing markets, and emerging technological challenges. These include cutting costs at home, shifting production abroad, or upgrading domestic production. New production concepts have a role to play in all of these strategies. However, the ISI manufacturing innovation survey indicated that the majority of German manufacturers have chosen a cost oriented strategy to commercialize the potentials of new production concepts. Nearly 60 per cent of the adopters of new technical and organizational options followed the paradigm of cost reduction for keeping competitive. The success of this strategy despite increased productivity by the use of new production concepts was merely poor. Only two fifths of the establishments which deployed new production concepts with a strategy oriented primarily towards reducing costs improved their market position and thus enlarged their turnover. However, three-fifths were unable to do so. These results suggest that even if German manufacturers try to reduce costs, they will still find it difficult to maintain markets for the cheapest, most routine products.

Yet, if German firms try to follow their competitors, especially from low wage countries, in the field of price competition, they have to shape new production concepts to conform to this objective. For example, we saw that teamwork in a cost-oriented strategy is not associated with consistent qualifications and skills for all team members, enabling each member to fulfill all tasks occurring within the group. The decentralization of tasks is shaped in a way that dominantly saves personnel instead of organizing decision making to the highest levels of information, skill and experience. Additionally, the no-buffer principle is of high priority and deployed to save stocks as much as possible. Productivity in terms of value added per employee can be increased by such combined initiatives.

However, we found that an even greater productivity benefit from new production concepts can be obtained if they are implemented within a performance-enhancing framework. Moreover, from an employment policy viewpoint, it is even more preferable if enterprises located in Germany attempt to strengthen their performance with the new production concepts rather than use these concepts to reduce their costs. While a successful cost-oriented strategy is better in terms of employment effects than an unsuccessful performance oriented strategy, the best employment outcomes are found through the successful implementation of new production concepts with a performance-oriented business strategy.

We did find that relocation for reasons of cost appears to raise the average productivity of locations in Germany. However, this is usually because the unproductive production processes are merely being "rationalized" out of existence. The parts of production that remain in Germany are often not improved. At the same time, in other cases, we found that a consistent reorganization of the processes taking place at the original location in Germany improves productivity to such an extent as to exceed the cost advantages that might be achieved by a foreign relocation. There are also additional advantages of continued domestic production in terms of flexibility and innovative capability.

The option of performance-orientation as a strategy to deploy new production concepts offers the possibility of maintaining existing customers and developing new ones through offering a high quality product. This strategy emphasizes quality circles, continuous improvement processes and ISO 9000 certification. These parts out of the whole spectrum of new production concepts obviously are most suitable to contribute to the envisaged strategy. The survey proved that these quality-oriented measures from the new production and organization concepts could reduce the rate of defectives. The differences between users and non-users of these new production concepts represent an improvement of 30 to 40 percent in rates of defectives.

A second type of performance orientation using new production concepts is targeted towards product innovation. If the company strategy is geared to innovation and technology leadership, then the ISI results show a clear, positive connection between the use of coordinated elements of new production concepts and innovation performance. This applies to the question whether new products can be brought to market maturity as well as to the speed with which product innovations can be pushed. Process innovations like simultaneous engineering, interdepartmental development teams, R&D cooperation with suppliers and customers, and continuous improvement processes are essential contributors to innovative products. New production concepts embedded in this kind of management strategy offer the enterprises the possibility to improve their innovative ability and to venture into market sectors which are not dominated by cost competition alone and moreover provide possibilities for growth.

A third attempt to differentiate the own products from those of the competitors in terms of performance is to offer not only investment goods to the customer but comprehensive solutions to each customer's problems. By following this strategy the manufacturer provides the goods with high value supplementary services. Because services are under a lower pressure from global price competition, a manufacturer can gain not only competitive advantage but also higher margins through combining product and services in a joint package. Teleservice is one opportunity for the producers of machinery to realize this strategy. The survey showed that this kind of reshaping of production to an innovative service orientation meet with increasing acceptance from customers. To reduce costly downtimes of the machines the customers of the machine producers appreciate what teleservice offers. Despite a higher machine price it is more effective for them to buy from a service-providing vendor in Germany.

Last but not least a performance-oriented strategy could try to gain competitive advantages by offering a "green" product. The survey indicated that a small share of German establishments is pursuing this route by implementing new production concepts like environment-audits. This measure for an environmental conservation is not only realized if the economic situation of the company is good. Environment-audits are obviously no measures only for good times. Companies that assess their profit situation less strongly than others seem to see the environment-audit as a chance. The environment-audit is a hope-bearer for businesses in order to improve their profit situation. The realization of measures to the environmental conservation is promoted by the innovative self-image of a company.

Cost-orientation and performance-orientation of new production concepts need to be carefully distinguished from each other. The analysis has shown that users of new production concepts can implement equally labeled elements (for example teamwork), but the shaping of these elements differ fundamentally due to the strategic orientation. But even if management not only implements the same label but the same concept, the exploitation may differ due to the strategy. Only the kind of selection and combination of elements from the "tool box" of new production concepts may give some hints on the strategy behind. Therefore the confrontation of different diffusion rates of new production concepts in an international comparison is not sufficient to indicate differences in performance and competitiveness. With an inadequate strategy "modern" production structures may have completely been introduced without any positive effect on competitiveness. On the other hand a less "modern" industry can compete very well if the combination of strategy and technical or organizational structures fits together.

Overall, the book highlights the importance of strategy and the roles of different contexts that have to be considered when trying to successfully implement new production concepts. In this sense, it is clear that new production concepts will not, as if by magic, automatically produce desired results. There is no general "cookbook" recipe, despite the implication to this effect given in many popular treatments of lean production and other fashionable business concepts. Rather, it has to be recognized that different national, industry and firm-specific contexts and cultures greatly determine the implementation and outcomes of new production concepts.

13.3   Implications for Management, Policy and Research

What consequences follow out of the results detailed in the book? We suggest that there are implications for at least three principal groups: industrial management, policy makers, and the industrial and technology policy research community. These implications are explored in the following sections.

Industrial Management and Business Strategy

Effective, proactive, and intelligent business strategy is crucial to sustained success in industry and to eventual economic and employment outcomes. There is, of course, much more to business strategy than the introduction of new production concepts. Nonetheless, at times, firms have focused on production concepts such as lean manufacturing or reengineering without fully considering the applicability and integration of these concepts. The results, perhaps not surprisingly, have been disappointing.

So, how should companies proceed in the implementation of new technological and organizational concepts? The point has already been made that new production concepts should be introduced within the framework of a broader business strategy, preferably one based on maximizing performance. Beyond this, there is clearly a need for customization and the development of individual strategies appropriate to particular industries and firms. The reorientation of activities flowing from a strategy to introduce new production concepts should be seen as a process rather than a set of set of piecemeal actions. And, strategy design as well as implementation should be envisaged as a joint process, involving employees as well as managers (see also Dreher et al. 1995, Lay and Mies 1997). An underlying condition for these approaches to work well is what Fleig and Mies (1996) have identified as "Vertrauenskultur" or a culture of trust. Where new production concepts have diffused rapidly in German companies, strong trust relationships throughout the company are often evident – indeed, the best German companies have long worked this way, reinforced softly by the legal framework for participation and codetermination.

The success of engaging affected employees in the implementation of new concepts gives emphasis on the (sometimes never ending) process itself. As shown in other ISI research on reorganization, internal communication is the main tool for reorganization efforts (Kieser et al. 1997). The goal is an adjustment of the individual and subjective organization concepts of the employees, their understanding of the organization, and the commonly defined organizational structure as target of the change efforts.

If there is any general conclusion for business from these studies it is that jointly elaborated solutions using the principles of new production concepts serving sound corporate strategies have significant positive impacts on productivity and market performance. Involvement and participation of the employees proved to be an important advantage in adjusting and optimizing the solution. Their participation is gained in an open and communicative process. Formerly often considered as an obstacle

We also conclude that businesses should be cautious of following fashionable trends. The life cycles of new paradigms become shorter and shorter. "Paradigm hopping" means to leave a concept before benefits had time to emerge. It is necessary to have a great staying power and the ability to adapt new ideas to the individual frame conditions. The management has to find the proper balance between organizational conservatism and "trend hopping."

New production concepts are no substitute for strategic deficiencies. New production concepts can support, not replace strategies. The shaping of new production concepts has to follow strategy. An extensive exploitation of new technological and organizational concepts requires a conclusive shaping of all restructuring measures. Strategy is the guideline to prevent contradictory orientations in the realization of different elements of new production concepts.

Industrial and Technology Policy

Policymakers have paid greater attention to promoting the diffusion of new production concepts in the most recent decade, with objectives related to industrial competitiveness, regional development, and employment. What are the implications of the ISI manufacturing innovation study for public industrial and technology policy, for government agencies, industrial associations, and for technology transfer organizations?

One set of consequences relates to the determination of objectives. The ISI study reaffirms a conclusion that other studies have found – that the modernization of industry through the implementation of new production concepts does not create jobs per se. Modernization can serve different purposes. Modernization as a defense measure may limit job losses or at best maintain employment. However, if modernization is conducted within a strategy to improve the overall innovative performance of a company, there can be the potential to create new jobs (as noted in chapter 5).

We also found, through our assessment of promotion schemes in the new Länder, that stimulating the adoption of new production concepts by investment grants is not sufficient for helping to make progress in performance. Achieving and sustaining high levels of industrial performance requires a consistent set of innovations. New technologies or organizational concepts are only parts of this set. They have to be embedded and accompanied by new technology management knowledge, new qualifications of the users, new incentives for the workers and sometimes a completely new enterprise culture.

The ISI study suggested that "slowing down the speed of diffusion" sometimes can mean "speeding up the performance progress." A more carefully considered decision making process and an implementation with more participation of the workers needs time. However, the results are more valuable for the establishment. Technology policy should realize these interdependencies and try to develop instruments not only to speed adoption processes but to improve the quality of new solutions.

Finally, the ISI results recommend that policymakers aim to encourage firms to deploy new production concepts within a performance-enhancing framework. This may mean that additional or complementary measures are required, to promote quality, product innovation, supplier-customer linkages, training, marketing, and business strategy. Even if production is unavoidably relocated to other countries to seek lower costs, continued efforts are needed to upgrade remaining elements of domestic production.

Research and Directions for Future Studies

A variety of research quantitative and qualitative methodologies are needed to develop a balanced assessment of industrial competitiveness and the diffusion and contribution of new organizational and technological concepts. Among these methodologies, comprehensive empirical studies of technological and organizational diffusion have an invaluable place. Ideally, such studies should be done on an ongoing basis – at least every few years, to ensure that up-to-date information is available as needs, conditions, and technologies evolve. ISI has been able to conduct several empirical surveys in Germany, including the 1995 study on which this book is based and a new 1997 survey. Technology and organizational diffusion studies have also been conducted in most other advanced countries (see, for example, U.S. Department of Commerce 1989, Alderman and Fischer 1992, Rosenfeld 1992, Swamidass 1994, NUTEK 1996, Shapira and Rephann 1996, European Foundation for the Improvement of Living and Working Conditions 1997, IDC 1998).

The first generation diffusion studies tended to focus primarily on "hard" technologies, particularly in the fields of automated and computer-aided manufacturing technologies. More recently, a second generation of studies has developed, which examines the diffusion of organizational or "soft" approaches, in addition to machine based technologies. These studies also aim to link organizational and technological change with company performance and innovation capability. The ISI survey on which this book is based is representative of this second generation.

What are the insights from the ISI study that can inform and guide future diffusion studies? First, the findings from the ISI survey confirm the point that it is not sufficient to investigate only diffusion and adoption rates of new production concepts in order to arrive at an assessment of the effectiveness of industrial modernization. The presence or otherwise of new production concepts per se is not a meaningful indicator of competitive production structures. Obviously new production concepts can contribute to competing successfully on the world market. However, methodological links have to be employed to track the use of new production concepts and their impacts not only on intermediate outcomes (such as improved quality or productivity) but also on bottom-line business measures (for example, sales or profitability) as well as metrics of policy concern (particularly jobs and wages). Moreover, the full impact of these concepts can only be appraised in the framework of a firm's overall business, production, marketing, and innovation strategy. Diffusion studies, particularly if they wish to assess impacts from adoption, need to incorporate elements that assess underlying business strategies – and explore the links between these strategies, the use of new methods, and business results. A related theme that, to date, has received inadequate attention is the link between the use of new organizational and technical methods in production and the development and commercialization of product innovations. Similarly, future research needs to better assess the implications of new ecological product and production concepts as well as holistic benchmarking initiatives. Understanding industrial modernization in such comprehensive ways will stimulate opportunities to explore relevant new research issues and develop appropriate findings.

Second, there is a need to strengthen the comparative aspects of diffusion studies. Although there have been some attempts to conduct multi-country studies of diffusion, generally the experience with diffusion surveys has been bounded within national or sub-national jurisdictions. Yet, it is clear that the value of such studies is much enhanced if international comparisons and benchmarks are possible. This is most likely to occur at the European level, as current research into innovation in firms within the European Union is expanded to take on a greater organizational and production technology dimension. But it would be desirable to extend the reach of comparative studies to other countries, particularly the United States and Japan.

Finally, fresh ways have to be found to diffuse the results from studies of organizational and technological innovation to a broader section of the business and policymaking community. Too often, businesses have to learn the hard way about the advantages and disadvantages of particular new technologies and methods. It may be too much to expect significantly better awareness and planning by users. But certainly the intermediaries that advise firms – including public industrial service providers, private consultants and industrial associations – could be supported to use more systematic approaches and tools to guide firms in selecting new technologies and methods and linking these to appropriate business strategies. Similarly, improved information and policy evaluation systems and greater dialogue between researchers, business, and policymakers could assist the policymaking community in developing more comprehensive demand-driven business assistance measures (within which organization and technological innovation is supported), rather than narrow, supply-side technology promotion schemes.

Dodgson, M., and Bessant, J. (1996), *Effective Innovation Policy: A New Approach*, International Thomson Business Press, London.

Dreher, C., Fleig, J., Harnischfeger, M., and Klimmer, M. (1995), *Neue Produktionskonzepte in der deutschen Industrie: Bestandsaufnahme, Analyse und wirtschaftspolitische Implikationen*, Physica-Verlag, Heidelberg.

European Foundation for the Improvement of Living and Working Conditions (1997), *New Forms of Work Organisation, Can Europe Realise its Potential? Results of a Survey of Direct Employee Participation in Europe*, Office for Official Publications of the European Communities, Luxembourg.

Fleig, J. and Mies, C. (1996), Multidisziplin - Erfolgreiche Unternehmen in einem turbulenten Umfeld, *Die Mitbestimmung*, 6, pp. 47-50.

Keynes, J.M. (1936), *The General Theory of Employment, Interest and Money*, Preface, Macmillan, London.

Kieser, A., Hegele, C., and Fleig, J. (1997*), Herausforderungen annehmen - Unternehmen gestalten: Organisation und Management zwischen Beharrung und Umbruch*, RKW-Verlag, Eschborn.

IDC (1998), Nordic ICT-O Survey, Council of Nordic Ministers, Copenhagen, Denmark.

Lay, G. and Mies, C., (Eds.) (1997), *Erfolgreich reorganisieren. Unternehmenskonzepte aus der Praxis*, Springer, Berlin and Heidelberg.

NUTEK (1996), *Towards Flexible Organisations*, Swedish National Board for Industrial and Technical Development, Stockholm, Sweden.

Rosenfeld, S. (1992), *Competitive Manufacturing: New Strategies for Regional Development*, Center for Urban Policy Research, New Brunswick, NJ.

Shapira, P. (1996), Modernizing Small Manufacturers in the United States and Japan: Public Technology Infrastructures and Strategies. In: Teubal, M, Foray, D., Justman, M., and Zuscovitch, E. (Eds.), *Technological Infrastructure Policy (TIP): An International Perspective*, Kluwer Academic Publishers, Boston, MA.

Shapira, P., and Rephann, T. (1996), The Adoption of New Technology in West Virginia: Implications for Manufacturing Modernization Policies, *Environment and Planning C: Government and Policy*, 14, pp. 431-450.

Shapira, P., and Rephann, T. (1996), The Adoption of New Technology in West Virginia: Implications for Manufacturing Modernization Policies, *Environment and Planning C: Government and Policy*, 14, pp. 431-450.

Swamidass, P. (1994), *Technology on the Factory Floor II: Benchmarking Manufacturing Technology Use in the United States*, The Manufacturing Institute, Washington, DC.

U.S. Department of Commerce (1989), *Current Industrial Reports, Manufacturing Technology Use, 1989*, Bureau of the Census, United States Government Printing Office, Washington, DC.

Appendix: Study Methodology

Steffen Kinkel and Martin Dreher

A.1   Introduction

The investment goods producing sector – which includes mechanical and electrical engineering, automobile production, metalworking, instruments, computers, and aerospace – is a critical component of Germany's economy, employing more than one-half of Germany's industrial workforce. Since the mid 1980s, the Fraunhofer Institute for Systems and Innovation Research (ISI) has conducted written surveys of firms in this sector, focusing on the adoption of new production techniques and organizational concepts and their economic, social and human resource impacts. ISI's surveys have been supplemented by company interviews and case studies to provide a comprehensive picture of the modernization strategies of manufacturers in the investment goods sector.

Investment goods sector surveys were undertaken by ISI in 1987 and 1990 in the "old" federal Länder (states) of West Germany as part of the evaluations of public programs to promote new technologies and methods in industry. In 1993, a comparable survey was undertaken in the "new" federal states of the former East Germany. In fall 1995, ISI conducted a further written survey, covering manufacturing establishments in the investment goods producing sector in both East and West Germany. This survey examined the diffusion of new techniques and production concepts, changes in personnel conditions, and new models of working hours and pay. Financial and other performance data were also collected, to allow analysis on the relationships between the use of new production concepts and economic outcomes. A total of 1,305 manufacturing establishments completed the 1995 survey. Results from the latest survey have been reported back to firms and industrial groups through targeted publications and newsletters, as well as made available through research studies and analyses. The 1995 survey also provides the common underlying database for the chapters in this book.

This appendix discusses the methodology used in the 1995 survey, including information on the parent population, the development of a survey database, the selection of a sample, the representativeness of the survey, and its administration.

A.2 Survey Objectives and Target Industries

(1) Survey Objectives

In the second half of 1995, ISI was commission to carry out a survey, using a written questionnaire, of manufacturers in the investment goods producing sector by the German Ministry for Education, Science, Research and Technology (BMBF). The Ministry sought information to help evaluate its program to promote computer-integrated manufacturing (CIM) in the new federal states. Non-participating firms, as well as firms that had participated in the program, were included in the survey. At its own expense, ISI extended the survey to include establishments in the investment goods sector in the old federal states, so as to cover all parts of Germany and allow comparisons between East and West German firms. ISI also added questions dealing with company production structures and the assessment of performance and competitiveness.

The final questionnaire asked manufacturers to provide information in the following categories:

- Industry, product range, product structure, and type of production
- Adoption and development of innovations in production operations
- Sales and profitability
- Supplier and customer structure
- Employment, wages and salaries
- Activities to increase competitiveness
- Targets and measures used in the manufacturing process
- Adoption of new technologies and techniques
- Application of new organizational elements
- Work organization, team work, task integration, and working time procedures
- Management of present technical-organizational projects
- Specific questions on the effects of CIM promotion (for participating East German plants).

A copy of the original questionnaire (in German) can be obtained from the Innovations in Production Group at the Fraunhofer Institute for Systems and Innovation Research.

(2)   Target Industries

The manufacturers targeted by the survey were establishments in the investment goods producing sector in Germany with 20 and more employees. In addition, those establishments with less than 20 employees were included in the survey if they formed part of a group of companies with 20 and more employees. The industries that comprised the investment goods producing sector are defined in Table A-1. This table also provides each industry's European Community General Industrial Classification code.

Respondents were asked to provide data at the establishment level. Where establishments are part of single-unit plants, this is relatively straightforward. For establishments in multi-unit plants, respondents were asked to estimate answers at the establishment level if specific data were not available.

Table A-1   Definition of the investment goods sector used in the ISI survey

European Community General Industrial Classification[a]	Industry Group
311-315, 319	**Metal fabrication,** including forging, stamping, secondary processing of metals, coating, sheet metal products, metal appliances and metal office furniture, excluding tools and finished metal goods
316	**Tools and finished metal goods**
321-328	**Mechanical engineering,** including industrial machinery and machine tools
33	**Office machinery and data-processing equipment**
341-348	**Electrical engineering,** including electrical, electronic, and telecommunications equipment
351-353	**Motor vehicles and parts,** including accessories
361-364	**Other transportation equipment,** including aerospace, railway equipment, and shipbuilding
371-374	**Instruments and precision optics**

[a]Converted from German SYPRO industrial classification codes (see Statistisches Bundesamt 1979).

A.3   Address Sources and Sample Design

The following data sources were used to identify establishment addresses:

- *ABC-Industriedatenbank*. After adjustments for industry and employment size, 12,520 addresses were accessible from the old federal states from this data source.

- *Institut für Wirtschaftsforschung Halle (IWH)*. The addresses of 4,161 establishments in the new federal states were available here, including lists of manufacturers outside of the investment goods industry and addresses of investment goods manufacturers with less than 20 employees.

- *CIM participants*. Addresses of 358 East German manufacturers participating in the CIM promotion program in the new federal states were obtained from project operators. These addresses were incorporated into sample database after comparison with the addresses supplied by IWH, to avoid duplication.

A structured design was used to select a sample of establishments from the above data sources. The total number of establishments that could be contacted was limited to about 8,000. A stratified design was used under which manufacturers were selected, without duplication, from a random sample of (West German) plants in the *ABC-Industriedatenbank*, all East German plants producing investment goods that could be identified from the IWH data, and all East German factories participating in the CIM promotion program, based on program operator's records. In all, a total of 8,077 establishments were included in this sample frame.

This stratified design resulted in a sample in which East German plants and East German CIM participants were over-represented. However, this procedure was necessary to arrive at a large enough sample for the evaluation project and to carry out statistically significant analyses of the East German manufactures. The consequences of this bias on the results of the study are explained in Section 5.3.

A.4   Response Rate and Structure of the Data Base

(1)   Response Rate

Questionnaires were mailed in September 1995 to the investment good sector establishments included in the sample frame. At the end of October 1995, a reminder letter was sent. This was sent to all 215 of the manufacturers that had participated in the CIM program but had not yet responded to the survey. A letter was also sent to a random selection of the remaining non-participating plants. In total, 5,998 reminder letters were mailed.

During the response period, the sample was adjusted to account for manufacturers found to have gone out of business, moved without a forwarding address, or been inaccurately classed within the investment goods sector. These 'losses' resulted in 927 establishments being excluded from the sample base. The effective sample thus contained 7,150 establishments (see Table A-2). By mid-December 1995, 1,305 usable questionnaires had been received from this effective sample, resulting in a response rate of 18.3 percent. The sample and effective response is schematically represented in Figure A-1. Although not complete, the ISI survey response rate is comparable with similar industry surveys, which often obtain response rates of between 15 to 25 percent.

Table A-2   Selection and adjustment of the establishment sample, ISI survey

Establishment location	Initial sample (establishments contacted)	Losses (through bankruptcy, etc.)	Effective sample (adjusted for losses)
Old Länder	{ 7,719	905 }	5,142
Non-participating New Länder[a]			1,672
Participating New Länder[a]	358	22	336
Total	8,077	927	7,150

[a]In the BMBF CIM program

(2)   Data Processing

Data collection and checking for consistency of the survey data was completed in spring 1996. To ensure completeness and consistency, some corrections and supplements were made to the survey responses. Information on industry branch and the number of employees were ascertained and added to the database, if these data were originally missing. (For mechanical engineering, detailed four-digit industry codes could be ascertained; for the other industries, only two-digit industry codes could be assured.) In addition, with the help of filter and control variables, inconsistent data were identified and removed from the database, for example where non-logical answers were given to questions on the use of technology. A non-response analysis was not carried out.

Figure A-1   Composition of responses

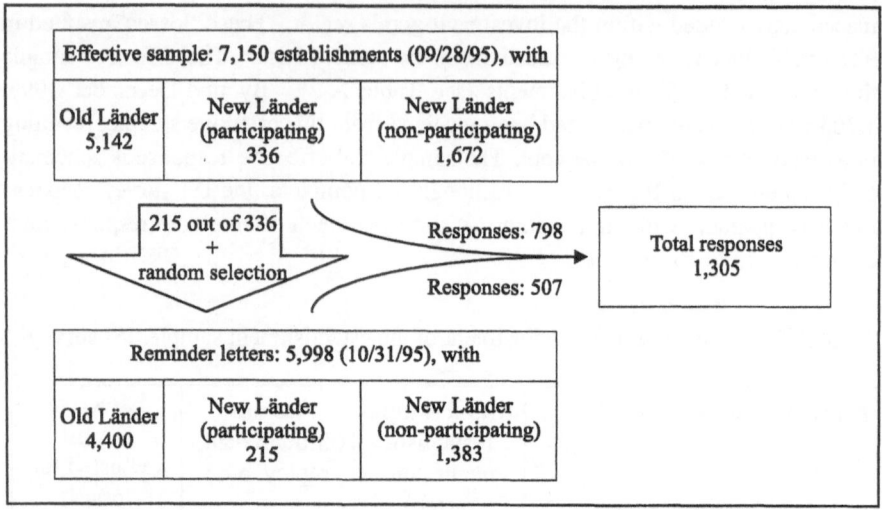

A.5   Characteristics of Survey Respondents and Comparison with Parent Industry Population

This section examines the characteristics of the establishments that responded to the survey. A comparison of the respondents is made with data for the parent population of investment goods manufacturers in Germany. The statistics for the parent population are taken from official statistics (see Statistisches Bundesamt 1992; Statistisches Bundesamt 1995). We use this data to examine respondent characteristics by regional location, establishment employment size, and industry branch. The survey design deliberately focused on over-sampling East German CIM program participants. To adjust for this and to compensate for other biases, a weighting scheme is used in the in the final survey analysis. The weighting scheme is described in section 6 of this chapter.

Compared with the parent population, the survey shows a clear preponderance of manufacturers from East Germany, while West German units are relatively under-represented. Establishments located in the new Länder returned 558 of the 1,305 usable questionnaires, or 43 percent of all survey responses. Yet, East German plants comprise only just over 12 percent of all German investment goods manufacturers. Based on their share of Germany's investment goods industry, East German plants were nearly five times more likely than those in West Germany to return questionnaires. Conversely, investment goods manufacturers from the old Länder of West Germany are relatively under-represented among survey respondents.

The much higher response rate from new Länder manufacturers reflects the special focus in the survey on CIM promotion program participants located in East Germany. ISI contacted all East German plants participating in the BMBF CIM program and sought a high response rate in order to conduct a program evaluation. Compared with the parent population, there is a significant over-representation in the database of CIM participant plants from the new federal states for this very reason (see Figure A-2). The objective of the promotional program was to accelerate the introduction and use of technologies and organizational procedures for computer-integrated production. It is reasonable to assume that the plants participating in this program differ from non-participants in the parent population, at least in their attitudes towards technological investment and organization change. To avoid misinterpretation of the results, the over-representation of CIM participants is a major factor for which adjustments are made in the weighting scheme used in the analysis of the survey results.

Figure A-2   Representiveness of survey respondents, by participation in the BMBF CIM promotion program

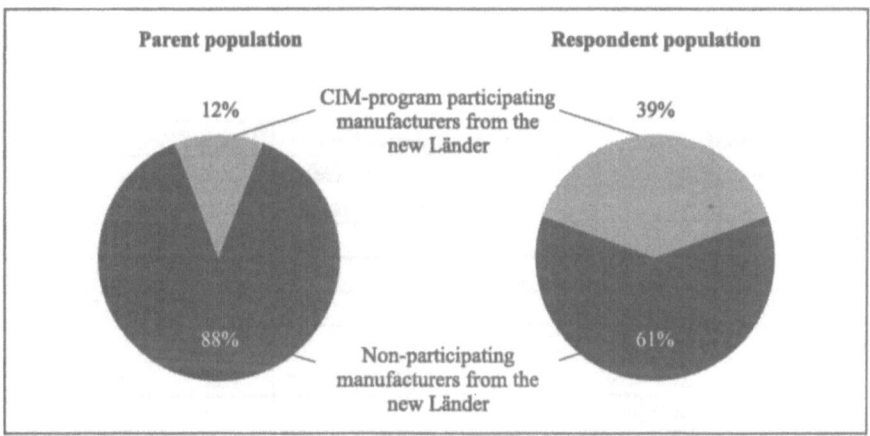

In terms of the industrial structure of the respondents, mechanical engineering establishments, which form nearly one-half of the survey database, are clearly over-represented compared with their share in the parent population of 30 percent (see Table A-3). Manufacturers involved in the motor vehicle industry are most under-represented in the survey database; they have a share of only 5 percent as opposed to their one-third share in the parent population. The remaining capital goods producing industries are all slightly under-represented in the database by comparison with parent population.

Table A-3   Survey respondents and parent population, by regional location and industry

Industries	Respondents (%)			Parent Population (%)		
	Old Länder	New Länder	Germany	Old Länder	New Länder	Germany
Mechanical engineering	49.9	45.3	48.0	30.3	29.1	30.1
Electrical engineering	16.3	17.9	17.0	20.0	19.5	19.9
Motor vehicles and parts	5.2	5.0	5.1	12.7	16.8	13.2
Finished metal products	10.8	9.0	10.0	12.9	9.5	12.4
Metal fabrication	7.1	15.2	10.6	8.6	16.2	9.6
Instruments and precision optics	5.2	3.9	4.7	6.5	3.6	6.1
Other investment goods	5.4	3.6	4.6	9.2	5.2	8.7
Total	100.0	100.0	100.0	100.0	100.0	100.0
Establishments (N)	747	558	1,305	20,277	2,880	23,157

Table A-4   Survey respondents and parent population, by regional location and employment size

Establishment Employment Size	Respondents (%)			Parent Population (%)		
	Old Länder	New Länder	Germany	Old Länder	New Länder	Germany
1-19	5.6	11.1	8.0	7.8	8.9	7.9
20-49	28.9	34.8	31.4	39.5	42.3	39.9
50-99	23.2	22.8	23.0	22.7	23.3	22.7
100-199	17.9	18.3	18.1	13.9	13.2	13.8
200-499	15.5	10.4	13.3	10.0	7.8	9.7
500-999	5.0	1.8	3.6	3.5	2.7	3.4
Over 1,000	3.9	0.9	2.6	2.6	1.9	2.6
Total	100.0	100.0	100.0	100.0	100.0	100.0
Establishments (N)	747	558	1,305	20,277	2,880	23,157

Differences in the employment size structure between the database and the parent population are smaller than those evident in the regional and industry comparisons (Table A-4). Establishments with 20 to 49 employees represent about 30 percent of the database, whereas the corresponding share in the parent population amounts to just under 40 percent. These smaller manufacturers are therefore slightly under-represented in the database, although manufacturers in the next class, those with 50 to 99 employees are represented at about the right level in the database, compared with their share in the parent population of about one-quarter. Manufacturers with 100 to 499 employees are slightly over-represented in the database – in the survey, their share is over 30 percent, whereas these establishments comprise a quarter of the parent population. Larger establishments, with 500 and more employees, are correctly represented in the database.

A.6 Weighting Scheme and Transferability of the Survey Results to the Parent Population

As shown in the comparison of the database with the parent population, there are considerable differences in the regional, industry and employment size characteristics of the survey respondents. As already noted, the survey database is also over-represented with CIM participants. In order to properly interpret and generalize the survey data results, it is necessary to weight the survey database to compensate for this.

The weighting scheme that is used adjusts for CIM promotion and East/West location, drawing on relationships between the original sample selection and the parent population. Weights are applied to three categories of manufacturers, as follows:

Establishments from the Old Länder
Weight = share of the parent population/share of the database
= (20 277 / 23 157)/(747 / 1 305)
= 1.530

Non-participating establishments in the CIM program from the New Länder
Weight = share of the parent population/share of the database
= (2 522 / 23 157)/(342 / 1 305)
= 0.415

Participating establishments from the New Länder
Weight = share of the parent population/share of the database
= (358 / 23 157)/(216 / 1 305)
= 0.093

When these weighting factors are applied, the regional and CIM program characteristics in the adjusted survey database are comparable to those among the general population of plants in the sector. The weighting scheme also has indirect effects on the industry and employment size distributions found in the survey data base compared with the distribution in the general population. For example, the

180

fers significantly from the distribution in the parent population only for the category of 20 to 49 employees (Figure A-3). This result is similar to that reported in other comparable surveys (SFB 187 1991 – 1995; Schultz-Wild et al. 1989). For establishments from the new federal states, the weighted distribution is almost identical with the size distribution in the parent population. The industry match is also closer, although even after weighting there are differences in the shares of the mechanical engineering and motor vehicle industries when compared with the general population. These need to be kept in mind when interpreting the survey results.

For the results reported in this book, the analysis is based on the three weighting factors described above for establishments from the old Länder, for new Länder CIM program non-participants, and for new Länder CIM program participants. Generally, median values are reported for the weighted results, to avoid variations due to dispersed values that the use of an arithmetic mean could produce.

Figure A-3   Relative distribution of weighted database, by employment size

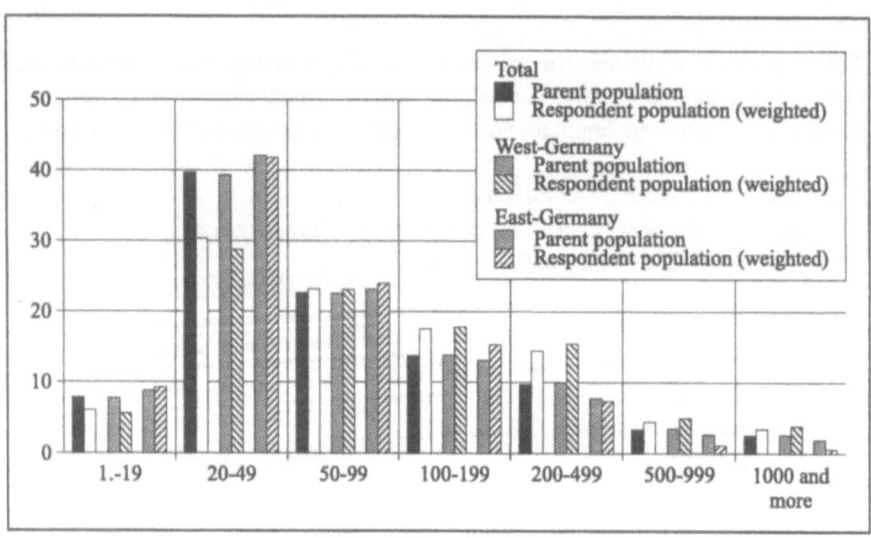

Figure A-4   Relative distribution of weighted database, by industry

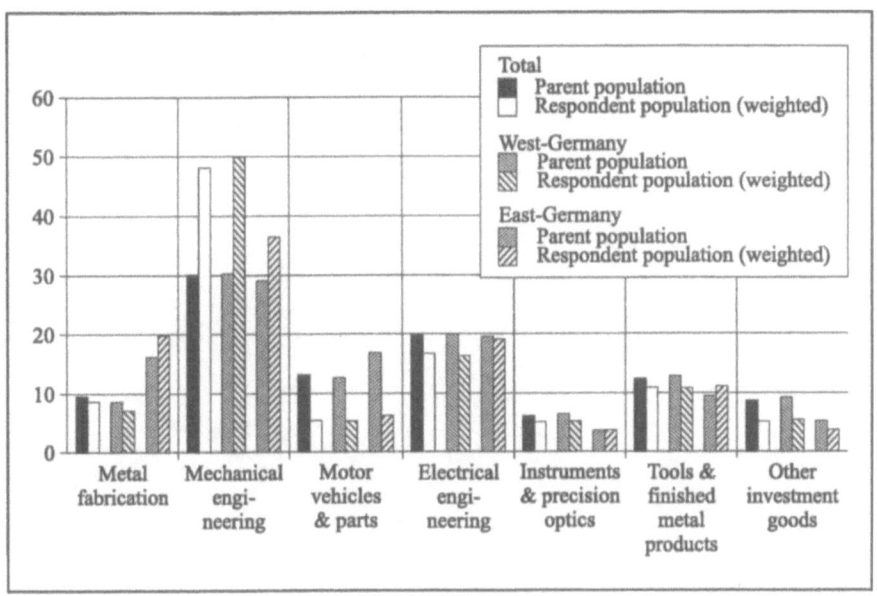

A.7   Bibliography

Schultz-Wild, R., Nuber, C., Rehberg, F., and Schmierl, K. (1989), *An der Schwelle zu CIM*, RKW, TÜV Rheinland, Eschborn, Köln.

Sonderforschungsbereich 187 der Ruhr-Universität Bochum (1991 - 1995), *Mitteilungen für den Maschinenbau*, No. 1-11 (August 1991 - September 1995).

Statistisches Bundesamt (1979), *Systematik der Wirtschaftszweige, Ausgabe 1979*, Fassung für die Statistik im produzierenden Gewerbe - SYPRO, Wiesbaden.

Statistisches Bundesamt (1992), Fachserie 4, Reihe 4.1.2, Wiesbaden.

Statistisches Bundesamt (1995), Fachserie 4, Reihen 4.1.1 und 4.1.4, Wiesbaden.

Contributors

Carsten Dreher graduated with a Diplom-Wirtschaftsingenieur (business administration and mechanical engineering) from the University of Karlsruhe and is qualified as a vocational trainer. He worked at the European Commission in the DGXII Monitor Program and joined the Fraunhofer Institute for Systems and Innovations Research in 1989. His research focuses on production technologies and organizational concepts for manufacturing systems, technological and organizational diffusion processes and their economic and societal impacts, and the evaluation of industrial and technology policies. Awarded a Ph.D. from the University of Karlsruhe in 1996, he now heads the Innovations in Production Group at the Fraunhofer Institute for Systems and Innovation Research. Dreher is the author of *Technologiepolitik und Technikdiffusion: Auswahl und Einsatz von Förderinstrumenten am Beispiel der Fertigungstechnik* – Technology politics and technology diffusion: Selection and use of methods to support production technology (Baden-Baden, Nomos, 1995).

Martin Dreher is now a researcher at the Institute for Industrial Production of the University of Karlsruhe. His previous work at the Fraunhofer Institute for Systems and Innovation Research focused on the design, implementation, and statistical data analysis of surveys about innovative production and organization concepts. He holds a diploma in industrial engineering from the University of Karlsruhe.

Jürgen Fleig is deputy head of the Innovations in Production Group at the Fraunhofer Institute for Systems and Innovation Research, where he undertakes research on production organization, production planning and ecological economics. He was previously a manager with a Karlsruhe software and consulting company, concerned with organizational and technical reorganization in the area of MRP-systems and manufacturing. Fleig holds a Diplom-Wirtschaftsingenieur (economic engineering) from the University of Karlsruhe and a doctorate in business administration from the European Business School at Oestrich-Winkel.

Steffen Kinkel studied industrial engineering at the University of Karlsruhe, specializing in corporate planning. He has been a researcher in the Innovations in Production Group at the Fraunhofer Institute for Systems and Innovation Research since 1996. His research focuses on the management and evaluation of innovative production and organization concepts, as well as problems in the management of globalization and the relocation of industrial activities.

Gunter Lay founded the Innovations in Production Group at the Fraunhofer Institute for Systems and Innovation Research, where he has worked since 1978. His research concentrates on the diffusion of technical and organizational innovations in production, the economic and social impacts of production innovations, and the evaluation of innovation promotion policies. He studied business administration at the University of Mannheim and holds a doctorate from the University of Kassel. In 1996, Lay was a visiting professor at the University of Grenoble. Among recent publications, Lay edited *Strukturwandel in der ostdeutschen Investitionsgüterindustrie* - Structural Change in the East German Capital Goods Industry (Physica Verlag Heidelberg, 1995) and co-edited, with Claudia Mies, *Erfolgreich reorganisieren* - Successful Reorganization (Springer Verlag Heidelberg, 1997). He currently co-ordinates ISI's ongoing survey on innovations in production in the German investment goods sector.

Thomas Michler is with the Technology Consultation Office of the German Trade Union Association, Rhineland-Pfalz. Previously, he worked with the Fraunhofer Institute for Systems and Innovation Research, focusing on use of new organizational and technological concepts in industry, the economic and ergonomic assessment of new production approaches, and the evaluation of technology policies. He holds a diploma in mechanical engineering from the Technical University of Munich.

Claudia Mies is a researcher with the Deutsche Telecom Research Institute. During her time spent at the Fraunhofer Institute for Systems and Innovation Research she worked on projects that examined changes in production structures, working time, personnel systems and remuneration. She holds a Master in Administrative Science from the University of Constance, Germany, and has also studied at Rotterdam, Netherlands and Warwick, UK.

Philip Shapira is Associate Professor in the School of Public Policy at Georgia Institute of Technology, Atlanta, and a visiting researcher at the Fraunhofer Institute for Systems and Innovation Research. He teaches and conducts research on the adoption and diffusion of new manufacturing technology and on the design, implementation, and evaluation of policies and programs for innovation, technology development, and technology transfer. Shapira served as a Fellow with the Office of Technology Assessment of the United States Congress and currently directs the

Georgia Tech Policy Project on Industrial Modernization. He holds a Ph.D. in city planning from the University of California, Berkeley.

Jürgen Wengel is deputy head of the Innovations in Production Group at the Fraunhofer Institute for Systems and Innovation Research. His research specialties include the analysis and evaluation of technology policies in the field of production technology at national and European levels, the diffusion characteristics of these technologies, technology transfer, and the impact of techno-organizational changes on firms, workforce and economies. He has served as a national expert with the European Commission, and managed national and European co-operative technology analysis and evaluation projects. Wengel studied at Göttingen University and the Hochschule for Verwaltungswissenschaften (administration science), Speyer, and holds a Diplom-Sozialwirt.

Werner Wallmeier studied Sociology at the University of Bamberg, specializing in empirical research methods, quantitative statistics and organization sociology. Since 1996, he has been a research assistant at the Fraunhofer Institute for Systems and Innovation Research. He mainly works on statistical data analysis, the design and implementation of surveys dealing with innovative production and organizational concepts.

The Innovations in Production Group at the Fraunhofer Institute for Systems and Innovation Research

The Innovations in Production Group at the Fraunhofer Institute for Systems and Innovation Research in Karlsruhe, Germany, conducts policy and industry-oriented research on enterprise competitiveness, new organizational concepts for production and logistics, computer aided management techniques, and the use of modern manufacturing methods and machines. A major theme is the assessment of the challenges and options to Germany as a location for industry; the group also conducts studies with European and international comparative dimensions and provides guidance and recommendations on shaping the innovation process for enterprises, associations and policy-makers. The group's work covers the following aspects:

Technology diffusion and adoption
- Description, analysis and foresight of the development of technology and organization in production
- Diffusion analyses in national and international comparison by sectors, regions, and types of firms
- Analysis of determinants of adoption decisions, derivation of diffusion potentials

Techno-organizational options and impacts
- Analysis of the economic, social and environmental impacts of techno-organizational alternatives
- Development and application of instruments for technology and organization management, e. g. extended viability analysis
- Supporting firms in individual reorganization projects

Public policy assessment
- Design and evaluation of public programs for the promotion of new production concepts and technologies
- International comparison of the aims, instruments and results of public promotion measures

For further information, please contact Dr. Carsten Dreher (cd@isi.fhg.de).

TECHNOLOGY, INNOVATION and POLICY

Series of the Fraunhofer Institute
for Systems and Innovation Research (ISI)

Springer and the environment

At Springer we firmly believe that an international science publisher has a special obligation to the environment, and our corporate policies consistently reflect this conviction.
We also expect our business partners – paper mills, printers, packaging manufacturers, etc. – to commit themselves to using materials and production processes that do not harm the environment. The paper in this book is made from low- or no-chlorine pulp and is acid free, in conformance with international standards for paper permanency.

Springer